Computer Mapping in Education, Research, and Medicine

HARVARD UNIVERSITY

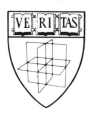

Laboratory for Computer Graphics and Spatial Analysis
520 Gund Hall, 48 Quincy Street, Cambridge, Mass. 02138

Contents

Computer-Based Instructional System For Teaching Thematic Mapping

by Paul R. Baumann

PROBLEM ENVIRONMENT

This computer-based instructional system has two primary objectives: (1) to introduce students to the decision-making processes associated with developing thematic maps, and (2) to induce students to use thematic maps as a tool for geographic research and as a means of spatial communication. As one of the few common denominators in geography, the map can be used in most geography courses, but often students are not given the opportunity to learn how to make or how to use maps. The main problem appears to be insufficient time to make maps. To produce maps using traditional techniques, students often spend a considerable amount of time and effort with drawing or "coloring in" procedures — generally considered "busywork" by students — and very little time learning how to formulate their decisions regarding the development of their maps. Also, students only have time to create one or possibly two maps in a course; consequently, they find it difficult to study the impact of various decisions on their maps or to utilize the map as a research tool or as a means of communication.

This instructional system uses the computer to draw maps rapidly and in quantity for students. With this computer system, much of the "busywork" is eliminated, allowing students to concentrate their efforts on the decision-making aspects of map construction.

In addition, students can use this instructional system to construct maps for the purpose of researching a topic or as a means of communicating ideas spatially. Again, the time-consuming procedures involved in drawing maps plus the failure to understand the decision-making processes curtail the proper construction of maps by students in many geography courses. Once students possess some comprehension of how to develop a map, they can use this computerized instructional system to create maps on many various subjects of their interest with relative ease and in a short time. By removing the time-consuming, laborious procedures associated with map construction, this system permits students to become involved with the map as a tool for geographic research and as a way to communicate ideas and information spatially.

THE COMPUTER MAPPING SYSTEM

Data Banks

This computer-based instructional system consists of a series of interlinking modules: data banks, a retrieval-transgeneration program, and two mapping programs (Figure 1). Each data bank contains data on a select number of key variables within a particular geographic area and the coordinate package needed to construct the different types of maps for the geographic area. Two data banks already exist, each containing around 55 variables. One data bank is on New York State with its 62 counties as subdivisions, and is used in an introductory geography course to instruct students in the decision-making aspects of mapping. The vast majority of the students in this course are from New York State. Working with a familiar area helps the students to associate patterns on their maps to other phenomena. The second data bank is on the urbanized area of San Antonio, Texas, with its 143 census tracts, and is utilized in an urban geography course to help students study the intra-urban spatial patterns of a typical large American city.

These data banks contain a limited number of variables because of logistic problems in making raw data in printed form available to each student. Key variables, mainly in the absolute form, have been selected since the transgeneration option of the retrieval-transgeneration program permits new variables to be formulated out of existing variables in the data bank. For example, from the variables — number of houses and land area — in the San Antonio data bank, housing density data can be derived for each census tract. Thus, from a few key variables thousands of new variables can be formulated.

The ease with which a data bank can be constructed gives the system its greatest flexibility in terms of course use. An instructor can design a data bank on any topic or area for which adequate data exist. Therefore, a data bank can be developed on any particular area or topic that an instructor wants to deal with in a course. An instructor's manual has been developed for this system explaining in detail how to build a data bank.[1] Properly supervised, a part-time work-study student can gather and code all the information needed for a data bank in one semester.

Retrieval-Transgeneration Program

The retrieval-transgeneration program forms the initial program in the system. Its main purpose is to retrieve data from the data bank. For each student map, a parameter information deck is developed containing various specifications relating to the construction of a map. These decks are read by this program in order to select the appropriate variables from the data bank. The data plus the parameter decks are placed on temporary disk files based on map types. At this point, the next phase in the system is ready to be executed.

This retrieval-transgeneration program also allows one to construct new variables based on variables already existing in

---

[1]P. R. Baumann, *Introductory Manual on Thematic Mapping: Instructor's Manual*, Project COMPUTe (Hanover, New Hampshire: Dartmouth College, 1976).

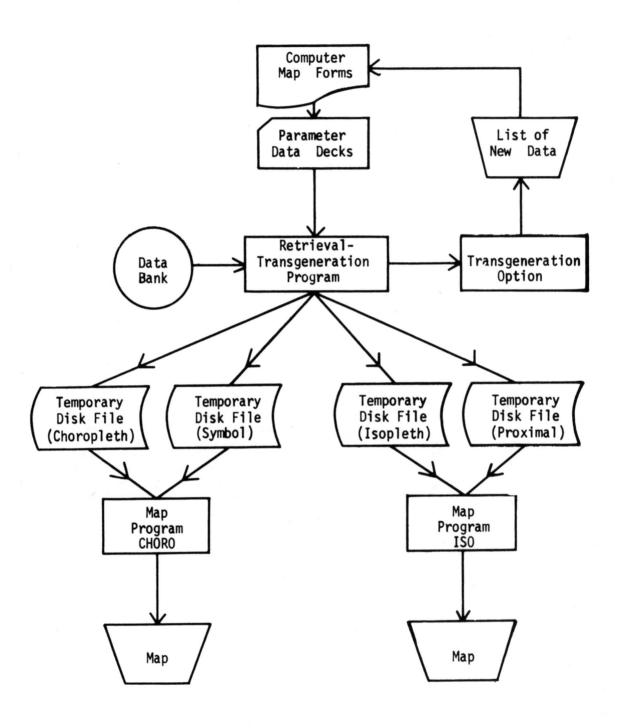

Figure 1.

the data bank. This option permits addition, subtraction, division, multiplication, and involution of one variable or constant to another variable. Other mathematical operations can be added easily if the instructor so desires. Only one operation can be performed with each transgeneration step. Presently, this system allows three transgeneration steps but the number can be expanded. This procedure is designed along the same approach as the Biomedical (BMD) transgeneration system.[2]

To illustrate the use of the transgeneration option, let us say that a student wishes to make a map showing the percentage of Negroes but finds that no such variable exists in the data bank. However, total population and Negro population are available. In the first transgeneration step, the total population is divided into the Negro population and the results are stored under a temporary identification number. Next, the results are multiplied by 100. Two mathematical operations are used; thus, two transgeneration steps are required. Since a new variable is being formed, a data list must be made available before any meaningful decisions can be made concerning the map. Therefore, in terms of output for transformed data, the system produces either a data list, a map based on either default options or specified parameters, or both a list and a map.

Mapping Programs

Once the retrieval-transgeneration phase is completed, the two map programs may be executed. These programs produce line printer-type maps as output. Although other output devices such as digital pen plotters, cathode-ray tubes (CRTs), electrostatic printers, and computer output-on-microfilm (COM) units make more visually appealing and more precise maps than the line printer, these devices are not present at many computer facilities; whereas the line-printer forms one of the standard output devices available at almost every computer installation. Also, the line printer can produce a large number of maps relatively fast in comparison to some of these other devices.

The one mapping program creates choropleth and symbol-type maps and is a modified version of Morton W. Scripter's CMAP, renamed CHORO for this system.[3] The other mapping program, ISO, forms isopleth and proximal-type maps and was developed specifically for this system.[4] Both programs employ the scan line approach used in CMAP to create map outlines and map cosmetics. A scan line contains the required coordinate data to produce a single map row by the line printer. With each line printer row being processed independently, only the coordinate data needed to produce a single row on the map must be stored in the computer's memory at any one time, eliminating any dependency on large and more expensive computers.

This system is capable of producing either one or hundreds of maps in one operational run. The amount of time to produce a map depends on the hardware and software systems, the type of map, and the size of the map. Isopleth maps generally take longer than other map types. If a large operational run which might require a considerable amount of time is anticipated, the run can be done in several phases or divided into several small runs.

The State University College at Oneonta possesses a Burroughs B 4700, a medium-size computer, with 200 KB (core bytes) or 400 KD (decimal digits) of memory. All three programs used in this system are written in FORTRAN IV and do not employ any special software or hardware features which might limit their use to the Burroughs systems. The size of the three programs ranges from 36 KB to 68 KB. At this size, the system can handle up to 500 data values per map. If the number of data values were reduced to 100, which can be accomplished by changing the size of certain variables in the DIMENSION statements, the size of the largest program can be decreased by 68 KB to 49KB.

INSTRUCTIONAL PROCEDURES

Before making any maps, either with or without the aid of the computer, students must be introduced to the importance and the role of maps. They must know how and when to use maps. To provide students with a certain level of map appreciation and understanding, a series of lectures is given on maps and map essentials. Also, a student manual, developed under a Project COMPUTe grant, is used to complement these lectures and to illustrate for students how to use the computerized aspects of this system.[5] Although drafting processes and techniques are discussed, the major emphasis in the lectures and the manual is on decision making, especially in the area of map content. Most students are not likely to become professional cartographers, but many of them are likely to be using and interpreting maps at various times in their lives. Therefore, they need to understand how different decisions influence a map. Students interested in pursuing map making in greater depth are encouraged to take a regular cartography course.

During a laboratory discussion session, students receive computer map forms (Figure 2) and a booklet which contains an outline map of the study area and its subdivisions, tables of the data available in the data bank, and the needed information to code the map forms. A computer-map form must be completed for each map. The information on the form is keypunched by a trained student to form parameter decks. It is not desirable to have each student keypunching his/her own material because of the number of potential errors generated by inexperienced keypunchers. Also, the time and effort to train students to do keypunching would distract them from the main objectives of this instructional system. Once the completed forms are returned to the instructor, the turn-around time is generally 24 hours. Students may submit as many forms at one time as they wish; however, they must recognize that the number often governs the length of the turn-around time.

In filling out the computer map forms, students face many decisions relating to the contents of their maps. First and most important, they must select their topic and decide if they are making their maps for the purpose of geographic research or to communicate a spatial theme. These decisions set the path for subsequent decisions. Next, they must search the data bank to find out if any of the variables can be used to produce their maps. Consequently, they must decide how to relate data to topic. Once a variable is selected, the next decision concerns

[2]W. J. Dixon, ed., **BMD Biomedical Computer Programs**, University of California Publications in Automated Computation No. 2 (Berkeley: University of California Press, 1968).

[3]M. W. Scripter, "Choropleth Maps on Small Digital Computers," **Proceedings** of the Association of American Geographers (Washington, D.C., 1969), pp. 133-136.

[4]P. R. Baumann, "ISO: A FORTRAN IV Program for Generating Isopleth Maps on Small Computers," **Computers and Geosciences**, 1978, Vol. 4, pp. 1-10.

[5]P. R. Baumann, **Introductory Manual on Thematic Mapping: Student Manual**, Project COMPUTe (Hanover, New Hampshire: Dartmouth College, 1976).

State University College
Oneonta, New York
Department of Geography

Item Code: ___ ___
[3 - 6]

Name: _
7 - 36

Trngen. Output: _
37

Map Type _
38

Transgeneration:

	X_k	Code	X_i	$X_{j \text{ or } c}$
TRNGEN	_ _ _	_ _	_ _ _	_ _ _
TRNGEN	_ _ _	_ _	_ _ _	_ _ _
TRNGEN	_ _ _	_ _	_ _ _	_ _ _
1 - 6	[7 - 10]	[11 - 12]	[13 - 16]	[17 - 26]

MAP PARAMETERS
Number of Levels: (minimum 2, maximum 10) _ _
[1 - 5]

Minimum Value: _ _ _ _ _ _ _ _ _
[11 - 20]

Maximum Value: _ _ _ _ _ _ _ _ _
[21 - 30]

RANGES

1	2	3	4	5
_ _ _ _	_ _ _	_ _ _	_ _ _	_ _ _
6	7	8	9	10
_ _ _ _	_ _ _	_ _ _	_ _ _	_ _ _
[1 - 10]	[11 - 20]	[21 - 30]	[31 - 40]	[41 - 50]

SYMBOLS

1	2	3	4	5	6	7	8	9	10
_	_	_	_	_	_	_	_	_	_
[1 - 2]	[3 - 4]	[5 - 6]	[7 - 8]	[9 - 10]	[11 - 12]	[13 - 14]	[15 - 16]	[17 - 18]	[19 - 20]

TEXT
Title: _
1 - 60

Source: _
1 - 60

999999 _
999999

Figure 2.

map type. Students must decide which map type can best represent their topic and data. After determining a map type, they must make some parameter decisions, such as the number of map levels, minimum and maximum data values, ranges or intervals of map classes, and symbol patterns. These decisions affect the content of the map and are very important. Finally, the title, source, and scale are determined. Students are taught that a title should be short in length and clear and accurate in content. The form limits the length of the title. To complete this form, students must make decisions relating to their maps, which is the primary goal of this instructional system.

RESULTS AND APPLICATIONS

As indicated previously, this system is being utilized in an introductory geography course and an urban geography course. Students in these courses have produced thousands of maps using this system, and, from their general reactions, they have attained a good understanding and appreciation of the decison-making role in map making. In comparing these classes with earlier classes in which students had to draw and color all their maps, the attitude of students toward map making is quite different. No longer does the instructor encounter the common grumbling of map making being "busywork." With this system students produce an average of seven to eight maps in a two-week period and they generally spend less time making these maps than they would producing one map under the traditional techniques. Now the moans and groans by the students are related to the difficult task of making decisions. However, the students are more curious about the patterns on their maps and spend more time developing hypotheses and testing theories to explain these patterns. Also, due to the ease of producing the maps, the students enjoy making maps of different variables so that they can compare the spatial relationships of variables.

In addition to the normal application of this system, students have made other uses of it. Some students have used the system to make maps for other classes. Several students have developed small atlases on special topics such as population, housing, transportation, and welfare. A few students have covered their maps with acetate and then have used marking pens to put other information on the maps. Usually the information is linear type data such as roads and railroads. Some students employed the system as a planning tool. For example, a map was made by one student showing what census tracts in San Antonio were saturated in terms of housing density and what tracts had growth potential. Several students have carried on experiments relating to decision making. As an example, one student made twelve maps of the same variable. Everything was constant on each map except the class intervals. When completed, the distribution of the variable differed drastically from one map to another. This type of experiment raises the question in the student's mind as to which map best represents the actual condition.

Most students taking the urban geography course are interested in the field of planning. This instructional system not only introduces them to the map as a powerful research tool and communication device for planning, but also enlightens them to the role of computer mapping and geo-information systems as planning tools. A number of these students take a course from the author entitled "Computer Mapping and Geo-Information Systems." In this course, this system is used as a model to illustrate how a geo-information system might be developed.

AVAILABILITY

The three programs used by this system are available at the nominal charge of $40 from Project COMPUTe, Dartmouth College. Figure 3 represents sample output from these programs. Eventually these programs will be available through CONDUIT, Box 388, Iowa City, Iowa 52240. In addition to the programs, the student manual and the instructor's manual can also be acquired from Project COMPUTe.

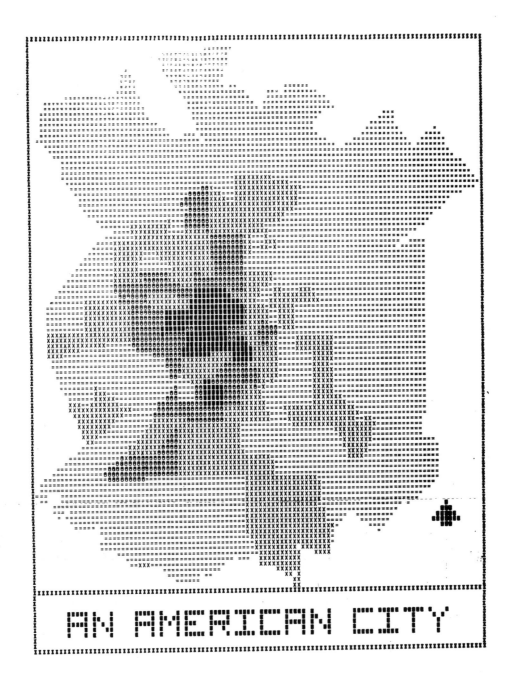

FOREIGN STOCK: MEXICAN
URBANIZED AREA OF SAN ANTONIO, TEXAS, 1970
U.S. CENSUS OF POPULATION AND HOUSING, CENSUS TRACTS, 1970

SCALE I--------I
MILES 0   2

JOHN DOE

ABSOLUTE VALUE RANGE APPLYING TO EACH LEVEL
MINIMUM   0.00   420.00   1260.00   2520.00
MAXIMUM   420.00   1260.00   2520.00   4200.00

LEVEL   1   2   3   4

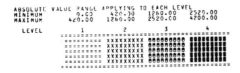

GEOGRAPHY   PROGRAM

DEPARTMENT OF GEOGRAPHY
STATE UNIVERSITY OF NEW YORK AT ONEONTA

SYSTEM DEVELOPED BY PAUL R. BAUMANN

Figure 3.

Use of Synagraphic Computer Mapping in Geoecology

by Elgene Box

THE STUDIES

Spatial patterns, such as species distributions and habitat topography, are often central to ecological studies. The term geoecology refers to spatial aspects of the ecology of large regions, including the entire world. Because of its ability to produce maps quickly and relatively inexpensively from irregularly spaced site data, SYMAP can be an extremely useful tool in both geoecological and smaller-scale ecological studies. The ecological studies at the University of North Carolina (UNC) in which SYMAP was used dealt almost exclusively with the entire world and involved modeling, mapping, and quantitative assessment of various environmental parameters and related biological phenomena. Among these were climatic factors, such as temperature, precipitation, evaporation, and solar radiation; plant productivity and energetics: plant forms and vegetation structure; and related ecological concepts such as the length of the growing season.

A central theme in these studies was the mapping and geographic assessment of plant productivity, which is the basis of ecological energetics and structure, and is one of the basic parameters of forestry, agriculture, range management, biogeochemistry, and (responsible) regional planning. Assessment of plant productivity (more properly called primary productivity) often involves correlations between plant production and various environmental factors, which provide a means of estimating productivity where measurements are not available.

The specific cartographic need for these studies was a convenient, reproducible, inexpensive means of displaying the various ecological parameters on large world maps. For world studies, several hundred data-sites represent a bare minimum requirement. The cartographic method employed must meet the following requirements:

1. It must be able to produce contour maps from irregularly spaced data, performing the necessary spatial interpolation.
2. It must be able to produce maps large enough to show world-scale detail.
3. It must provide for mathematical and other manipulation of data at execution time.
4. It must be inexpensive enough to use in a university environment.
5. It must produce results in a form which can be analyzed quantitatively by computer.

SYMAP, of course, meets all these requirements.

After constructing a world base-map of the land areas and a package of related otolegends, we first used SYMAP to produce a computerized simulation of an earlier world map of annual net primary productivity (Lieth 1964). Datum-sites were chosen as needed to reproduce the earlier pattern. Datum values consisted only of the indices of the contour levels, i.e.,

the integers one through seven. This first map was presented in Innsbruck and was called the "Innsbruck Productivity Map" (Lieth 1972). It represents the land portion of the map shown in Figure 1.

The idea of using the computer to analyze the finished SYMAP maps followed very quickly. Among the analytical and manipulative capabilities desired were planimetry, digitization of boundaries, and map overlaying.

A series of para-cartographic programs to perform these operations on SYMAP maps was outlined by Box (1975b) and developed. The most recent versions of the programs are documented by Box (1978b) and in individual user's manuals. The program for planimetry (MAPCOUNT, see also Box 1975b, 1978d) proved to be the most immediately useful of these programs and represents an integral part of this documentation.

Resource Requirement

The primary resources required for the studies were the SYMAP program (bought from the Harvard Laboratory), the computerized world base-map, various ecological data-bases, the analytical programs (primarily MAPCOUNT), and various means of coordinate conversion. At the time the studies began in the early 1970's, computerized base-maps were not as readily available as they are now. Since student manpower was available, the base-map (A-OUTLINE) was traced by UNC students from a large wall map produced by Rand-McNally. This map is based on a new projection by Robinson (1974) which represents a compromise between equal areas and correct shapes. It is very good for visually oriented displays since the distortion in polar regions of equal-area projections is greatly reduced.

Coordinate-conversion and map-reading analytical programs were written by the author primarily at the Nuclear Research Center (Kernforschungsanlage, KFA) in Julich, Germany, where the author was active during 1973-74. Although use of MAPCOUNT and its sister programs is reasonably inexpensive, development of the programs required funding not available at UNC. The four map-reading programs were first documented in an in-house paper for the KFA (Box 1975a).

Problems Encountered

Several potential problems were encountered while developing the system. These involved:

1. projection mathematics and coordinate conversion;
2. refinement of the base-map and of MAPCOUNT in order to produce more accurate areas;

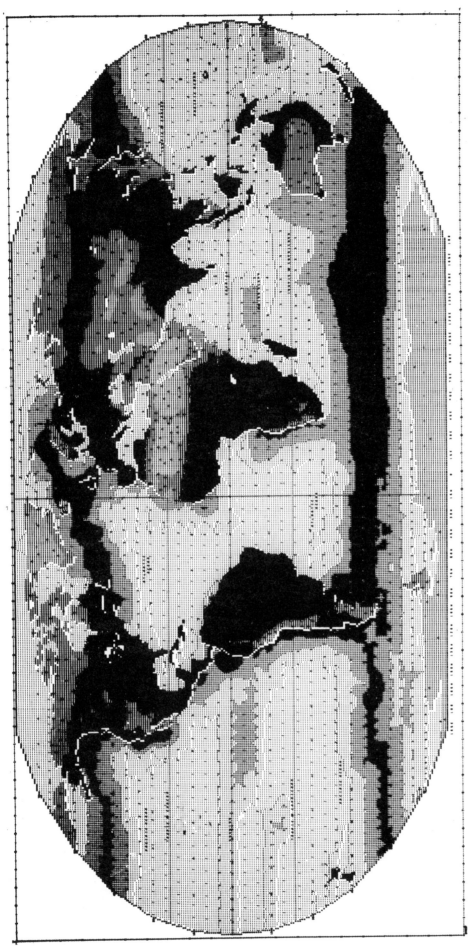

Figure 1.

3. photography of the finished maps;

4. sharp boundaries between disjunct regions, such as between land and sea;

5. the SYMAP upper limit on the number of data-sites;

6. the SYMAP requirement that B-DATA POINTS and E-VALUES be provided on separate cards.

The first problem was solved by the publication in 1974 of the mathematical specifications of the Robinson projection, and by construction of a general FLEXIN routine called FLEXPROJ to read geographic coordinates (for all cylindrical and pseudo-cylindrical projections) and convert them to map coordinates at execution time. The second problem was solved in two steps. MAPCOUNT was originally written to count only entire print-fields but was converted to a continuous coordinate system in late 1974. At this time a correction procedure was also added to MAPCOUNT (version 2.0), by which true areas of counted regions could be provided by the user in order to correct (standardize) the computed thematic results.

Photography of the large maps remains problematic but can be accomplished by at least two methods. At the KFA the SYMAP output could be written directly onto microfilm in map format by means of a locally developed routine which aligned the vertical strips of multi-page maps. This type of photography is illustrated by Figures 1 and 6. At UNC the maps are printed and hung against a white background for photography with high-resolution film. Dark, even prints are required, which computer operators sometimes find inconvenient. This process generally requires several trials but eventually produces more satisfactory results (Figures 2, 3, 4, 5, 7 and 8).

The problem of sharp boundaries between land and sea was solved by the development of MAPMERGE, a relatively simple map-reading program for overprinting up to five SYMAP maps (Figure 1 and Box 1978a). MAPMERGE also permitted us to include more data-points on one map.

The problem of data input on separate cards was solved by means of a very simple routine (applied through FLEXIN) which reads both B-DATA POINTS and E-VALUES during B-DATA POINTS processing, stores the E-VALUES on temporary disk space, and recalls the values during E-VALUES processing. This temporary storage of E-VALUES is included also in the above-mentioned routine, FLEXPROJ.

APPLICATIONS AND GRAPHICS

The primary graphics produced were the large SYMAP world maps, especially those of plant productivity. From a cartographic standpoint some of the maps are quite similar, but scientifically they show a number of different phenomena.

Plant Productivity and Energetics

The terrestrial "Innsbruck Productivity Map" was joined by a companion contour map of the estimated actual net primary productivity (NPP) of the oceans (Lieth, Hsaio, and Van Wyck 1972). These two complementary maps were overprinted by MAPMERGE to produce the "Seattle Productivity Map" (Figure 1), so called because it was presented in Seattle (Lieth et al. 1972). The maps were produced from a total of about 2,600 productivity estimates (datum-sites). One can see immediately from Figure 1 that the geographic patterns of productivity on land and in the sea are quite different.

The next project, called the "Miami Model," involved the relation of net primary productivity to two of its major environmental determinants, temperature and precipitation. Correlation curves were constructed to relate NPP to each factor separately, and

then, reflecting the well-known ecological Law of the Minimum (Liebig's Law), the smaller of the two estimates of NPP was taken as the Miami Model estimate. The estimated values of NPP for 1,001 sites were mapped to produce the Miami Model map of what can be interpreted as one estimate of potential net primary productivity (Box, Lieth, and Wolaver 1971). A slightly improved version of this map (Miami Model 1a, with 1,230 sites, from Box 1978d) is shown here as Figure 2. This model and map proved to be our most useful, since temperature and precipitation data are readily available. All the maps produced were evaluated by means of the planimetry program MAPCOUNT in order to obtain quantitative global and regional estimates (see below). The original version of the Miami Model also served as the basis for a study of possible variations in net primary productivity resulting from changes in climate. Initial results were published by Lieth (1975a) and are discussed below.

As more data became available from studies of the International Biological Program and other sources, it became possible to explore not only net amounts but also the basic energetics of plant production in some detail. The first project in this direction involved production of a world map of energy fixation by natural vegetation, which depends not only on production levels but also varies slightly with type of vegetation. In order to determine vegetation type(s) at each site, the basic climatic limits for each of 13 biome-level vegetation types were determined and the site-distributions of these types were generated using the program ECOSIEVE (Box 1978c). Knowledge of net productivity, vegetation type, and mean energy content of the particular type of biomass (from Lieth 1975b) permitted construction of the so-called "Berlin Model" of average annual energy fixation of natural vegetation (Figure 3), presented in Berlin (Box 1976). The conversion of vegetation and energy-content values into values of energy fixation was performed within FLEXIN.

The Berlin map was subsequently compared with a SYMAP simulation of Geiger's (1965a) world map of estimated annual solar input in order to produce a map of estimated annual photosynthetic efficiency (Box 1977), shown here as Figure 4. Since the two maps involved different sets of data-sites the division of energy-fixation values by values of solar input to estimate photosynthetic efficiency was performed for each print position using the overlay program MAPMATH, another of MAPCOUNT's sister programs (Box 1978a). Figure 4 shows an interesting pattern of maximum photosynthetic efficiency in the humid tropics and secondary maxima in cool-temperate and maritime-subpolar areas. However, the map also illustrates one of the problems with photography of printed maps on which the print is not even across rows.

Other studies involving estimation of plant dark respiration and gross production, growing seasons, and actual evapotranspiration also yielded world maps which were quantified by MAPCOUNT. A summary of some of the geographic results and world patterns is given by Box (1978d).

Physical Environmental Factors

It was also very useful to produce maps of various factors of the physical environment which influence the biological phenomena being studied. World maps of certain aspects of temperature and precipitation were presented by Lieth and Box (1972) and by Box (1978c). One of the most interesting of these maps shows the average precipitation of the warmest month,

E. Box, Chapel Hill

0 .1 .25 .5 1.0 1.5 2.0 <   kg/m²·yr

Figure 2. Miami Model 1A: Net Primary Productivity

Figure 3. Annual Energy Fixation of Natural Vegetation

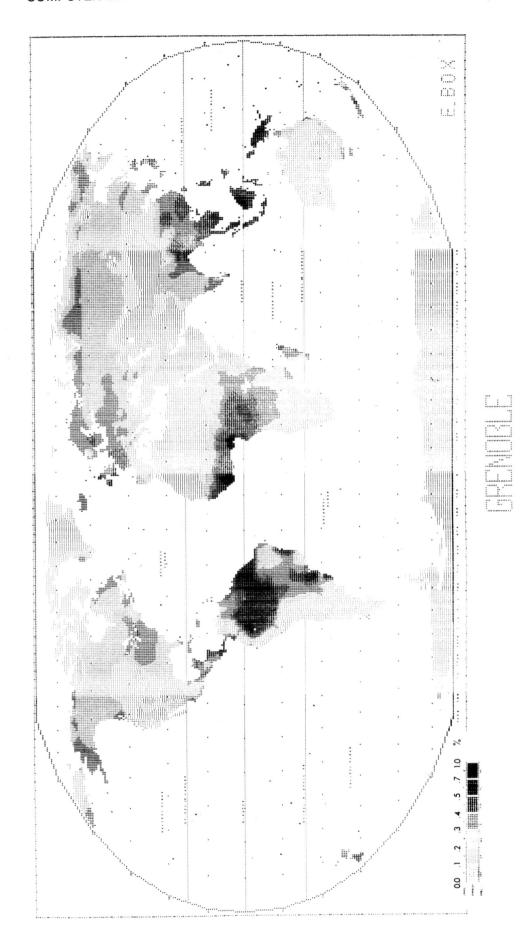

Figure 4. Annual Photosynthetic Efficiency

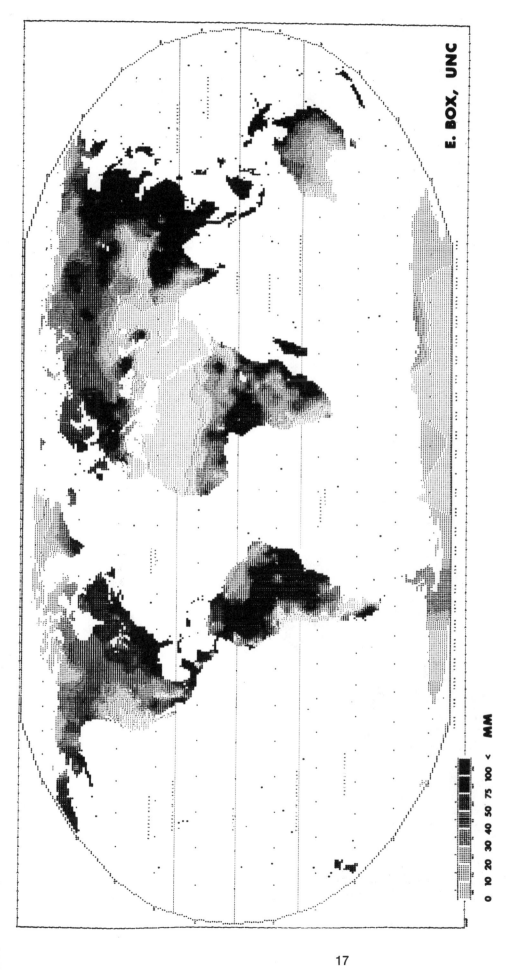

Figure 5. Pmtmax: Average Precipitation of Warmest Month

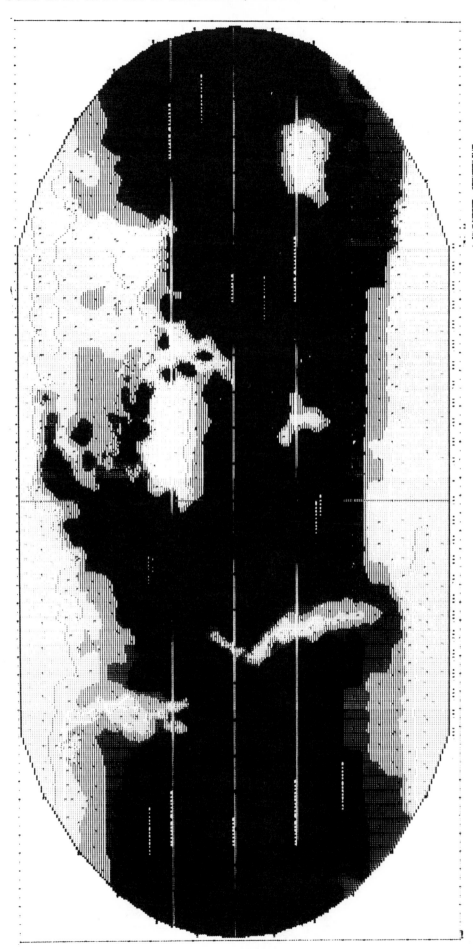

Figure 6.

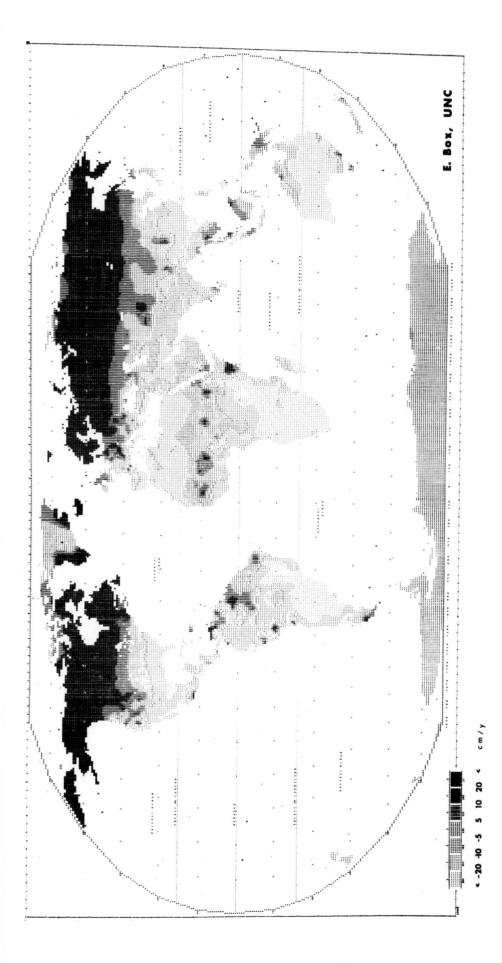

Figure 7. Thornthwaite PET minus Holdridge PET (Annual)

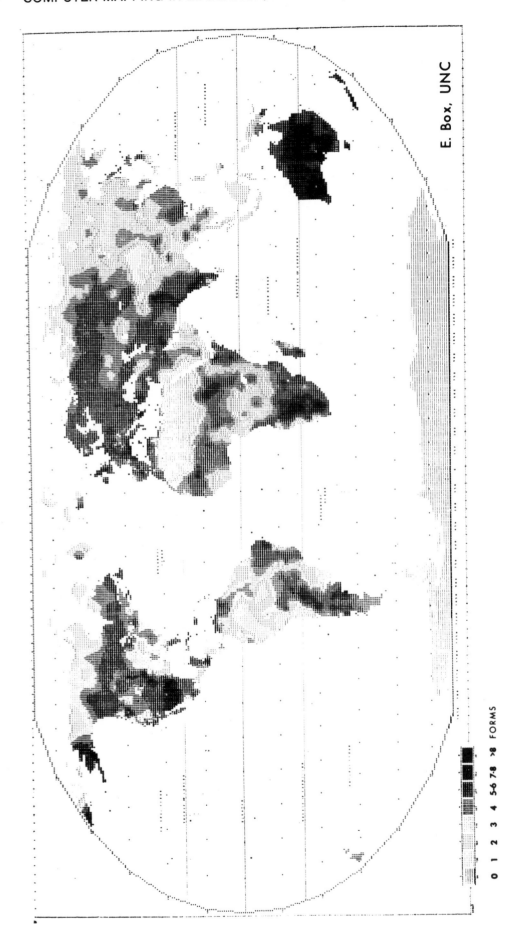

E. Box, UNC

0 1 2 3 4 5-6 7-8 >8 FORMS

Figure 8. Predicted Physiognomic Diversity of Dominant Vegetation

Figure 9.

	Terrestrial			Marine		
Latitudinal Belt	Area 10^6 km^2	NPP g/m^2	NPP 10^9 t	Area 10^6 km^2	NPP g/m^2	NPP 10^9 t
90-80° N	0.26	139	0.036	3.6	-	-
80-70° N	3.3	111	0.37	8.2	65	0.54
70-60° N	13.3	258	3.44	5.5	142	0.78
60-50° N	14.6	427	6.24	10.9	210	2.29
50-40° N	16.5	489	8.08	14.9	202	3.01
40-30° N	15.6	548	8.56	20.8	87	1.81
30-20° N	15.2	629	9.55	25.1	53	1.34
20-10° N	11.4	978	11.11	31.6	53	1.67
10-0° N	10.1	1688	17.08	34.2	73	2.51
10-0° S	10.5	1899	19.89	33.8	82	2.78
20-10° S	9.5	1245	11.79	33.5	78	2.61
30-20° S	9.3	744	6.94	31.0	69	2.13
40-30° S	4.1	712	2.91	32.4	94	3.04
50-40° S	0.9	716	0.68	30.8	221	6.81
60-50° S	0.2	200	0.04	25.3	319	8.08
70-60° S	2.0	42	0.08	16.8	143	2.41
80-70° S	8.7	26	0.23	2.8	73	0.20
90-80° S	3.9	1	0.003	0.0	-	-

The values were obtained by MAPCOUNT planimetry of the land and sea portions of the "Seattle Productivity Map" (Figure 1). The procedure is explained in more detail by Box (1978d).

Table 1. Estimated Actual Net Primary Productivity of the Earth's Land and Ocean Areas (Annual).

			Miami Model with $T \rightarrow T - 1°C$ $P \rightarrow P - 6\% P$			Miami Model with $T \rightarrow T + 1°C$ $P \rightarrow P + 6\% P$		
Latitude Belt	Land Area (10^6 km^2)	SMM NPP (t/km^2)	NPP (t/km^2)	$-\triangle$	$-\%$	NPP (t/km^2)	$+\triangle$	$+\%$
70°-60° N	13.3	387	355	32	3.0	426	39	11.0
60°-50° N	14.7	645	600	45	7.0	690	45	7.0
50°-40° N	16.5	804	756	48	6.0	849	45	5.6
40°-30° N	15.6	856	816	40	4.7	891	35	4.1
30°-20° N	15.2	748	720	24	3.2	771	23	3.1
20°-10° N	11.4	1042	1004	38	3.6	1086	44	4.2
10°-0° N	10.1	1797	1734	63	3.5	1853	56	3.1
10°-0° S	10.5	1964	1924	40	2.0	2002	38	1.9
20°-10° S	9.5	1402	1340	62	4.4	1469	67	4.8
30°-20° S	9.3	882	841	41	4.6	931	49	5.6
40°-30° S	4.1	856	815	41	4.8	904	48	5.6
Total (10^9 t)	149.4	127	121	6	4.7	132	5	4.0

SMM = Standard Miami Model map (Box, Lieth, and Wolaver 1971)
NPP = Annual Net Primary Productivity
T = mean annual temperature (°C)
P = average annual precipitation (mm)
$\triangle$ = difference between NPP of particular model and that of SMM
% = percent difference

Table 2. Estimated Variation of Annual Net Primary Productivity with Hypothetical Worldwide Changes in Temperature and Precipitation Levels.

Latitudinal Belt	Solar Input (kly)	Mean Energy Fixation (10^3 kcal/m^2/y)	Mean Phot. Efficiency (%)
90-80° N	65	0.48	0.074
80-70° N	65	0.61	0.093
70-60° N	78	1.97	0.25
60-50° N	91	2.58	0.28
50-40° N	114	2.46	0.22
40-30° N	142	2.34	0.16
30-20° N	160	2.19	0.14
20-10° N	164	3.55	0.22
10-0° N	151	5.72	0.38
10-0° S	150	6.22	0.41
20-10° S	156	4.63	0.30
30-20° S	150	2.95	0.20
40-30° S	131	2.49	0.19
50-40° S	103	2.20	0.21
60-50° S	81	2.09	0.26
70-60° S	66	0.14	0.02
80-70° S	75	0.10	0.01
90-80° S	85	0.0	0.0
Total Land	129	2.85	0.22

The values for solar input and energy fixation were obtained by MAPCOUNT planimetry of a simulation of Geiger's (1965b) solar radiation map (not shown) and of the "Berlin Model" map (Figure 3). The values for photosynthetic efficiency (column 4) were obtained by dividing energy fixation (column 3) by solar input (column 2).

Table 3. Estimated Average Annual Solar Input, Energy Fixation by Terrestrial Vegetation, and Photosynthetic Efficiency.

which proved to be a useful index of water stress during the growing season. This map is reproduced as Figure 5.

One of the most attractive maps produced is a simulation (Figure 6) of the color map by Geiger (1965b) of estimated annual effective evapotranspiration of the globe. The ghostly black-and-white SYMAP version dramatizes the differences between land and sea much better than does the color map. The land portion of this map was used also as the basis for a model relating annual net plant productivity to actual evapotranspiration (Lieth and Box 1972).

Potential evapotranspiration is one of the basic theoretical and practical tools of many climate-related studies since it represents the climatic potential water loss from any wet surface. This theoretical capacity of the air to absorb water is primarily a function of energy input, relative humidity, and air motion. It can be estimated in many ways. Two of the most widely used methods, both based on air temperature, are those of Holdridge (1959) and Thornthwaite and Mather (1957). Both have been tested in various parts of the world, but apparently a systematic, cartographic worldwide comparison had not been made. SYMAP (with FLEXIN) was used to compute both estimates at 1,225 sites and map the differences between the two estimates. The resulting map (Figure 7) shows clearly that the Holdridge estimate is, in general, much higher in the tropical areas and the Thornthwaite-Mather estimate much higher in temperate and polar areas. In order to have a single method which is reasonably accurate worldwide (except in windy areas), it does not seem without merit to consider averaging the two estimates.

The author has also used SYMAP with various, often rather complex, FLEXIN packages to model and display the results of a variety of other physical environmental parameters including the earth's solar input, soil types and their water-holding properties, lengths and threshold values of plant growing seasons, and other climatic and soil factors. Most of these maps have not yet been photographed or published.

Vegetation Types

The SYMAP proximal-mapping algorithm was found to be very useful in mapping both actual and predicted vegetation types. In one lengthy study (Box 1978c), world terrestrial vegetation was classified into 59 "ecophysiognomic" plant forms presumed to be physiognomic adaptations to particular climatic situations. Examples include the broad-leaved evergreen trees of areas with constantly favorable conditions, the broad-leaved winter-deciduous trees of humid climates with cold winters, the stem-succulents and microphyllous shrubs of warm semi-desert climates, and so on. For each plant type, environmental envelopes were estimated with respect to various aspects of annual temperature, precipitation, and potential evapotranspiration regimes. World distributions of the 59 types were then generated using the general screening program ECOSIEVE and 1,225 climatic data-sites. World patterns of both the climatic variables (e.g., Figure 5) and the predicted plant types were displayed as SYMAP proximal or contour maps. In addition, the numbers of plant types predicted at each site were displayed as a contour map in order to present an estimate of the world pattern of form diversity in vegetation (Figure 8). This diversity has important implications for environmental management as well as for general ecological structure and function. Comparing Figures 5 and 8, one can see that warm-season precipitation and form diversity in vegetation are inversely

related to a considerable degree. Photography of the vegetation maps is more problematic because of the need to distinguish individual symbols. A one-page section of the map of predicted distributions of narrow and needle-leaved tree types (as formation dominant or co-dominant forms) is shown in Figure 9. Maps of the predicted world patterns of particular plant-physiognomic characters are also planned.

Other Studies

SYMAP has also been used with excellent results for several other types of ecological studies by UNC personnel and their colleagues. Photographs are not available, but examples are listed in order to show the range of ecological applications.

One of the earliest studies involved the mapping of airborne industrial pollutants in the United States (Wolaver and Lieth 1972), an area in which SYMAP has found many uses. A series of maps for different pollutants was produced. Each map shows major sources and the geography of its effects on surrounding areas.

SYMAP was also used by Reader and Radford to map the arrival of spring and autumn in North Carolina and eventually across the entire eastern United States (Reader, Radford and Lieth 1974; also Lieth and Radford 1971). The arrival of the seasons was determined by the flowering of dogwood (*Cornus florida*) and redbud (*Cercis canadensis*) in spring, and by the leaf-coloration of red maple (*Acer rubrum*) and tulip poplar (*Liriodendron tulipfera*) in autumn. The study of spring's arrival was extended to the western states (using a species of lilac) by Caprio et al. (1974), who produced a map of the arrival of spring for the entire conterminous United States.

Two recent studies of animal behavior at UNC have employed SYMAP to present the results of spatial foraging patterns by small animals. Hall (in preparation) used otolegends to display paths traveled by turtles against a background of study-area topography based on E-VALUES representing site micro-elevation. Alho (1977) used SYMAP to map the foraging ranges of a population of mice in a forest habitat in order to determine areas covered and the effects of particular foraging territories on other members of the population.

Using the same basic methods for plant productivity as outlined above, but with a much larger number of data-sites in the United States, Sharpe (1975) produced excellent SYMAP maps of the net primary productivity of the United States as predicted by two climatic models.

At the Julich research center (KFA), Forstel et al. (1975) have used SYMAP to display the world pattern of the oxygen-18 isotope content of rainwater, which is an important parameter for studies of the earth's oxygen and water cycles involving naturally occurring isotope fractionations in photosynthesis and transpiration. Their final map was quantified using MAPCOUNT to obtain a latitudinal histogram relating standardized ^{18}O-content of rainwater to temperature.

RESULTS

Once produced, each map was stored on tape where it could be accessed easily for printing or for quantification by MAPCOUNT. MAPCOUNT planimetry consists of determining the areas covered by the different SYMAP levels and multiplying them by average thematic values for the individual levels (e.g., average productivity per unit area). Both areas and thematic quantities are summed over all levels to obtain totals for the entire map or for the subregion treated. If the true area of the region treated is known, it can be submitted to MAPCOUNT to correct the thematic sum for whatever error in area representation existed in the map and/or the counting procedure. The impact of the mapping and planimetry procedures on the particular problems is best treated by individual study.

World Plant Productivity and Energetics

The world maps of primary productivity and related energetics were quantified globally, by 10° latitudinal belts, and by large land masses. Although world estimates of primary productivity and energy fixation had been advanced by a few authors, never before had it been possible to attempt such detailed regional estimates.

Although general world patterns of actual primary productivity were perceived (Lieth 1964), planimetry of the land and sea maps (Figure 1) provided a first truly quantitative geography of estimated actual annual primary productivity. The results are summarized in Table 1 (from Box 1978d). For the land areas, total annual net production is estimated to be about 108 x 10^9 metric tons of dry matter, a figure which is in general agreement with most other estimates. The Northern Hemisphere produces about 60% of this total. Latitudinal mean values are seen to decrease poleward. The region between 20° N and 20° S is seen to produce about half of the global total.

The geographical pattern of ocean productivity, on the other hand, is quite different. The total annual marine production is less than half the terrestrial production despite a much larger marine area. The highly productive southern subpolar zone (50° S to 70° S) is seen to yield about one-third of the world's total marine production.

Possible quantitative changes in global NPP due to changes in global climate were studied using the climatically based Miami Model (Figure 2). By increasing or decreasing mean annual temperature by 1° or 2° C at each site (with an associated, hypothesized 6% increase or decrease in annual precipitation), it was possible to generate a series of modified Miami-Vodel maps to be quantified by MAPCOUNT. The estimated variations in NPP resulting from 1° modifications of global temperature are summarized in Table 2. It was found from this study that a uniform 1° C increase/decrease in global temperature could result in a 5% increase/decrease in total terrestrial (natural) NPP, the effect being greater poleward. The effect on agricultural productivity would be much greater, since agriculture relies mostly on certain end-products of plant production. A 1° C decrease in mean global temperature would eliminate agriculture from large areas.

Quantification of the energy-fixation (Figure 3) and solar-radiation maps permitted the construction of a similar geographical budget for production energetics (land areas only). The results are shown in Table 3. The total annual energy fixation by the terrestrial vegetation cover is seen to be about 425 x 10^{15} kcal, about half of which is fixed in the tropical zone between 20° N and 20° S. Latitudinal mean values are generally much higher toward the equator, but the latitudinal trend of latitudinal mean values also shows plateaus across both the northern and southern temperate zones. The trend across the northern temperate zone is even inverted. This is probably caused by the greater diversity of climate and vegetation types in these zones.

The values of mean photosynthetic efficiency in Table 3 were obtained by dividing the mean energy-fixation values by the corresponding solar-input values. These values yield an average of 0.22% for the photosynthetic efficiency of the entire terrestrial plant cover. Separate MAPCOUNT assessment of the map of estimated photosynthetic efficiency (Figure 4) yielded an estimate of 0.23% for the land areas. Efficiency is seen to be highest in the tropical lowlands, low in areas where water may be limiting; secondary maxima are in the subpolar belts (50° to 60°) of both hemispheres, where production is low but where solar input is also greatly reduced by prevailing cloudy conditions.

Similar quantification of the other productivity maps (Box 1978d) led to the following observations:

1. Total annual terrestrial dark respiration of plants amounts to about 133 x 10^9 metric tons, which is only slightly higher than the global estimate of potential net production predicted by the Miami Model. Total terrestrial gross production (defined to be the sum of net production and dark respiration) is estimated to be about 263 x 10^9 tons annually. Latitudinal means of dark respiration vary from less than 40% of gross production in polar areas to about 60% in the tropics.

2. Of the continents, South America shows the highest mean net productivity over its entire land area. Mostly, temperate Eurasia and subarid Australia show the lowest values.

3. The Miami and Montreal models of potential net primary productivity, as estimated from correlations with climatic variables (Lieth and Box 1972), show higher estimates of net production than does the map (Figure 1) of actual net production. The differences are greatest in the northern temperate zone, and seem to represent a good measure of man's reduction of the productive capacity of the land he occupies.

Vegetation Distribution and Structure

The vegetation maps were not quantified but were used instead to check the ability of the hypothesized environmental envelopes of various plant forms to provide a reasonable simulation of world vegetation distributions and structure. This geographic validation was complemented by a more formal comparison of predicted and actual vegetation at particular, well-documented sites. It was found that the model predicts actual vegetation in about 75% of the cases. Warm-season precipitation, as expressed in Figure 5, was found to be most closely correlated with form diversity of vegetation (Figure 8) but in a negative way. This was not unexpected, but the degree to which Figures 5 and 8 present mirror-images of each other was not expected and would not have been detected without producing the world maps.

Physical Environment

Only one example is discussed. In the course of evaluating the map of estimated annual solar input to the earth for use in estimating photosynthetic efficiency, it became apparent that there is a wide variation in various estimates of annual global solar radiation, from a low of about 2.1 x 10^{24} joules (Lieth 1975b) to a high of 3.2-3.8 x 10^{24} joules (Kouchkovsky 1977 and personal communication). This variation prompted us to examine the problem in more detail and to quantify our computerized simulation of Geiger's map. The resulting values for 10° latitudinal belts are included in Table 3.

The global total obtained for the simulation of the Geiger map was 2.72 x 10²⁴ joules per year, which is near the middle of the range of estimates and is in general agreement with the bulk of the other estimates. There is still considerable uncertainty in local values, however, especially in the polar areas of the map. The annual receipt of solar radiation on earth is one of the most basic physical parameters of our planet. As new data become available, they can be added to the SYMAP simulation, and it is hoped that this methodology may provide a convenient means of refining the estimate of solar input.

CONCLUSIONS AND HINDSIGHT

The potential usefulness of SYMAP and synagraphic mapping has already been documented in a wide variety of fields, quite apart from whatever the present paper may provide. SYMAP's usefulness in geoecology and in more localized ecological studies should be equally obvious, since so many ecological phenomena have important spatial components and implications. The potential of MAPCOUNT and computerized planimetry as a convenient tool for quantitative geographical assessments, using the medium of SYMAP maps, has been shown by MAPCOUNT's impact on world assessments of plant productivity and other geoecological phenomena.

In hindsight, two possible changes in the course of MAPCOUNT development might have been made. In order to compute area-related thematic totals using mapped data, one must have accurate thematic weights for the various SYMAP levels. These can be computed easily by averaging the E-VALUES falling in the respective intervals. It would be very simple to include this computation in the SYMAP job itself, either as a SYMAP elective or as a standard part of a FLEXIN routine.

The other potential change (which obviates the need for the first) involves adaptation of MAPCOUNT and its sister programs to read SYMAP's elective 21 output, which provides the numerical values interpolated for each print position. This would greatly increase the resolution of the map-reading programs.

BIBLIOGRAPHY

ALHO, C.J.R. 1977 "Spatial Distribution of *Peromyscus leucopus* in Different Habitats." Chapel Hill, North Carolina: University of North Carolina, Ph.D. thesis.

BOX, E. 1975a. "MAPCOUNT und andere Programme zur Herstellung, Weiterverarbeitung, und quantitativen Auswertung statistischer, vom Computer gedruckter, Landkarten." Unpublished documentation of computer programs developed at the Kernforschungsanlage Julich GmbH, Julich, Germany. 88 pp. plus appendices (in revision, 1978).

BOX, E. 1975b. "Quantitative Evaluation of Global Productivity Models Generated by Computers." *Primary Productivity of the Biosphere*. H. Lieth and R. H. Whittaker, eds. Springer-Verlag.

BOX, E. 1976. "Berlin Model Map of Average Annual Energy Fixation of Natural Vegetation." *Applications of Calorimetry in the Life Sciences*. Lamprecht and Schaarschmidt, eds. p. 328, Walter de Gruyter.

BOX, E. 1977. "Estimated Annual Photosynthetic Efficiency of

Terrestrial Vegetation." SYMAP world map presented at the European Seminar on Biological Solar Energy Conversion Systems, Grenoble, May 1977.

BOX, E. 1978a. "The MAPCOUNT Series of Programs for Analysis and Maniipulation of SYMAP Results."

BOX, E. 1978b. "Quantitative Geographic Analysis: A Summary (with Geoecologic Applications) of Para-Cartographic Programs Developed at the Julich Nuclear Research Center." Julich/Germany: Kernforschungsanlage Julich GmbH, Jul-Bericht (in preparation).

BOX, E. 1978c. "Ecoclimatic Determination of Terrestrial Vegetation Physiognomy." Chapel Hill, North Carolina: University of North Carolina, Ph.D. thesis.

BOX, E. 1978d. "Geographical Dimensions of Terrestrial Net and Gross Primary Productivity: A First Attempt." *Radiation and Environmental Biophysics* (in press).

BOX, E., H. LIETH, and T. WOLAVER. 1971. "Miami Model world map of annual net primary productivity predicted from temperature and precipitation." *Publ. in Climatology, 25:37-46; Human Ecology*, 1:303-32: and *Primary Productivity of the Biosphere*. Lieth and Whittaker, eds. Springer-Verlag.

CAPRIO, J., R. HOPP, and J. WILLIAMS. 1974. "Computer Mapping in Phenological Analysis." *Phenology and Seasonality Modeling*. H. Lieth, ed. Springer-Verlag.

FORSTEL, H., et al. 1975. "The World Pattern of Oxygen-18 in Rainwater and its Importance in Understanding the Biogeochemical Oxygen Cycle." *Isotope Ratios as Pollutant Source and Behaviour Indicators*. Vienna: International Atomic Energy Agency.

GEIGER, R. 1965a. "The Atmosphere of the Earth, map no. WA6: Annual Effective Evapotranspiration." Darmstadt: Justus Perthes.

GEIGER, R. 1965b. "The Atmosphere of the Earth, map no. WA1: Total Annual Short-Wave Radiation." Darmstadt: Justus Perthes.

HALL, S. (in preparation). Ph.D. thesis in zoology, University of North Carolina (Chapel Hill).

HOLDRIDGE, L. R. 1959. "Simple Method for Determining Potential Evapotranspiration from Temperature Data." *Science*, 130:572.

KOUCHKOVSKY, Y. DE. 1977. "Can the Efficiency of Photosynthesis be Increased?" Paper presented at the European Seminar on Biological Solar Energy Conversion Systems, Grenoble, May 1977.

LIETH H. 1964. "Versuch einer kartographischen Darstellung der Produktivitat der Pflanzendecke auf der Erde." *Geographisches Taschenbuch* 1964/65, pp. 72-80. Wiesbaden: Steiner-Verlag.

LIETH, H. 1972. "Uber die Primarproduktion der Pflanzendecke der Erde." Zeitschr. fur Angew. Botanik, 46:1-37.

LIETH, H. 1975a. "Possible Effects of Climate Changes on Natural Vegetation." *Atmospheric Quality and Climatic Change*. R. Kopec, ed. *Studies in Geography*. vol. 9. Chapel Hill: University of North Carolina.

LIETH, H. 1975b. "Primary Production of the Major Vegetation units of the World." *Primary Productivity of the Biosphere*. H. Lieth and P. H. Whittaker, eds. Springer-Verlag.

LIETH, H., and E. BOX. 1972. "Evapotranspiration and Primary Productivity; C. W. Thornthwaite Memorial Model." *Publications in Climatology*, 25:37-46. Elmer/New Jersey.

LIETH, H., et al. 1972. "Seattle Productivity Map of global actual annual net primary productivity." Presented at the 5th Genl. Assembly of the Intl. Biol. Pgm., Seattle, August 1972. Documented in: *Primary Productivity of the Biosphere*. Lieth and Whittaker, eds. Springer-Verlag.

LIETH, H., E. HSAIO, and P. VAN WYCK. 1972. "Oceans Productivity Map of annual marine net primary productivity." *Primary Productivity of the Biosphere*. Lieth and Whittaker, eds. Springer-Verlag.

LIETH, H., and J. S. RADFORD. 1971. "Phenology, resource management and synagraphic computer mapping." *BioScience*, 21:62-70.

READER, R., J. S. RADFORD, and H. LIETH. 1974. "Modeling Important Phytophenological Events in Eastern North America." *Phenology and Seasonality Modeling*. Lieth, ed. Springer-Verlag.

ROBINSON, A. H. 1974. "A new map projection: Its development and characteristics." *Internatl. Yearbook of Cartography*, 14:145-55.

SHARPE, D. M. 1975. "Methods of Assessing the Primary Production of Regions." *Primary Productivity of the Biosphere*. H. Lieth and R. H. Whittaker, eds. Springer-Verlag.

THORNTHWAITE, C. W., and J. R. MATHER. 1957. "Instructions and Tables for Computing Potential Evapotranspiration and the Water Balance." *Publications in Climatology*, 10(3):185-311. Elmer/New Jersey.

WOLAVER, T., and H. LIETH. 1972. "The Distribution of Natural and Anthropogenic Elements and Compounds in Precipitation across the U.S.: Theory and Quantitative Models." Chapel Hill, North Carolina: University of North Carolina Duplicating Shop. 75 pp. offset.

The Evolution of the South Carolina Coastal Mapping Program

by David J. Cowen, George Walters, Kenneth Feinberg, Michael Holland and Alfred Vang

PROBLEM STATEMENT

The emergence of the federal government into areas of environmental concern during the decade of the 60's has imposed an enormous burden on state governments to develop mechanisms for complying with the mass of technical and administrative requirements accompanying each new program. As one recent account suggests, "the primary task of state agencies in natural resource management is to move rapidly and efficiently to identify the resources that require management to see that they are protected. Information gathering is always subservient to these goals."[1] In order to meet these goals, new technical and administrative programs for gathering, organizing, and analyzing relevant information in a meaningful fashion have been created. However, "this process involves assembling a vast array of natural resources, land use, demographic, and other information relevant to the nature of the resources, the development pressures acting upon it, and the actions necessary to protect it."[2] Furthermore, collection and integration of these data must, in most cases, be achieved within very limited budgets, as well as within personnel and time constraints. When one recognizes that the federal government now has more than 130 programs requiring some form of analysis of land and water related data, the inevitable confusion and inefficiency that results become understandable. States often react with frustration since they believe that they have been delivered administrative and technical problems for which they are neither adequately compensated nor prepared. Subsequently, in order to receive their state's fair share of the taxpayers' money, they find themselves becoming unwillling customers of federal programs.

The purpose of this paper is to discuss how the state of South Carolina implemented an automated mapping system to help meet the requirements of one of these federal programs. Specifically, it discusses the development of the South Carolina Coastal Mapping Program. Although this program was designed to meet the needs of the Coastal Zone Management Act, the evolutionary process discussed should be representative of the experiences that any state becoming involved in automated geographical data handling would encounter. The paper presents a case history of an incremental approach to the conceptualization, design, construction and operationalization of a system in which a state university played a major role. Basically, the paper concludes that such an incremental approach is much more realistic than is the process advocated in the idealized model found in the literature of system development.

THE SOUTH CAROLINA EXAMPLE

The passage of the Coastal Zone Management Act of 1972 (PL 92-583) has singled out the fragile ecological system of the coastal zone for special environmental attention. As finally passed, the act relied upon the states to develop and establish administrative procedures to enforce a coastal zone management plan. Specifically, the Act mandates that the states involved designate: (1) coastal zone boundaries, (2) permissible uses, (3) areas of particular concern, (4) areas of preservation, (5) areas of priority uses, and (6) organizational structures.[3] The determination of the first five of these items requires the detailed analysis of a diverse set of geographical related data. As one of the states covered by the Act, South Carolina was forced to formulate a program for handling such information.

It has been estimated that South Carolina has approximately 500,000 acres of tidal marshland along its 200 miles of coast line. These marshes constitute one of the state's most prized resources. Rapid increases in tourism, second-home ownership and economic growth have demonstrated the need to plan for the orderly use of this resource. Concurrent with the recognition of this requirement was the demand for a mechanism which could both draw together the variety of information relating to coastal resources and present it in a manner conducive to planning and management activities. When the governor appointed the members of the Coastal Zone Planning and Management Council in 1973, he charged them with the responsibility for establishing such a mechanism, as well as developing a planning and management program for the state.

The initial proposal for a grant under Section 305 of the Coastal Zone Management Act placed a high priority on the development of an automated resource inventory system. In fact, about twenty-four percent of the 1974 grant of $198,000 was devoted to a computer system design in which "general inventory data were digitized, processed and stored."[4] As in any such program, the major assumption was that automation would offer an efficient, rapid, and cost-effective approach to data collection and analysis. The general concepts of computer cartography seemed to have been sufficiently developed to offer the promise of developing a dynamic system that would enable the state to rapidly update various data items, as well as supply specific information for management decisions.

Conceptually, the specific objective of the initial system was to update existing base maps in the coastal zone. Since it was the general feeling of the council that planning and management should proceed on a site-specific basis, the USGS 7 1/2-

[1]American Society of Planning Officials, *Information/Data Handling Requirements for Selected State Resources Management Programs* (Washington, D.C., U.S. Department of the Interior, 1975), p. 5.

[2]Ibid., pp. 506.

[3]Management of the Coastal Zone (PL 02-583), Oct. 1972 Title III Sec. 305.

[4]The South Carolina Coastal Zone Planning and Management Council, "An Application for a Program Development Grant to the Office of Coastal Environment" (March, 1974), p. 32.

minute quadrangle was selected as the standard base map for the program. Unfortunately, an analysis of the status of this map coverage for the ten counties selected for the program quickly revealed the inadequacy of those maps. Of 175 potential 7 1/2-minute quadrangles, only 101 had been published. One-third of those had not even been photo-revised since 1960. Furthermore, the only USGS coverage of the dynamic Myrtle Beach area was a 1937 15-minute quadrangle. Therefore, it was mandatory that a new system be designed that could encode natural and cultural features from recent source materials and then merge that data with the information from the USGS quadrangle to produce a revised base map at a scale of 1:24,000. Essentially, the initial concept necessitated a highly accurate digitizing and plotting system for cartographic display.

One of the most far-reaching consequences of the Act resulted from its philosophy regarding data collection. Specifically, the requirement to inventory land use and other data "should not be construed as requiring long-term continuing research base line study, but rather as providing the basic information of data critical to successful completion of a number of required management program elements . . . "[5] This interpretation, coinciding with President Nixon's "New Federalism" which permitted the states to act individually in their data collection and analysis efforts, seems to have undermined the development of accurate management plans in the coastal zone. In fact, it could be argued that extensive and standardized data collection efforts should have been mandated. Instead, the government's policy encouraged the states to use existing data sources. Therefore, the ability of states to develop adequate inventories became a function of already existing base-line data, such as that provided by the USGS National Mapping Program. Succinctly stated, the individual states were required to develop management plans aimed at the same goal, but commencing at very different points, with little assistance from the federal government regarding definitions, scales, or data bases to be employed. Consequently, and not surprisingly, thirty separate and largely incompatible plans have been emerging. Furthermore, it could be argued that the program, with such a worthy goal, is sorely underfunded. For example, the 1974 allotment for 305 grants was only $7.2 million; in 1975 that amount totaled only $9 million. Although the size of these grants if a function of the length of the coast and its population density, it is doubtful that the structure necessary to administer, collect and analyze the data would vary in direct proportion to these measures. Therefore, the system, operating in such a manner, necessarily penalizes the smaller states.

THE INITIAL SYSTEM DESIGN

The initial system that evolved was developed at the Charleston County Planning Office. The hardware system consisted of a Bendix Digitizer, a tape drive, a Calcomp 760 flatbed plotter and a Honeywell computer, all of which were utilized to perform other county functions. Low altitude, color infrared photography was used as the basic resource for updating the quadrangle sheets and for interpreting land use and wetland vegetation. Responsibility for the production of the wetlands inventories was given to the South Carolina Wildlife and Marine Resources Department, Division of Marine Resources Laboratory in Charleston. That group conducted extensive field surveys involving salinity sampling and vegetation mapping. The wetlands

inventory was superimposed on a mylar overlay using a zoom transfer scope. Although the existing quadrangle and the updated wetlands inventory formed the basic input in the system, other data, such as state-owned lands, existing land use, oyster and clam grounds, and archaeological sites, were to be added over time.

The initial data base concept assumed that a map was merely a set of lines. Each line on the source materials was designated a certain line type. This scheme included both linear features, such as roads and single line creeks, and boundaries around natural and political areas. For example, the line surrounding a brackish marsh could be subdivided into a number of boundaries, such as lowmarsh/brackish marsh, brackish marsh/upland, and fresh marsh/brackish marsh. Each of these boundary types was given two numerical codes that identified the areas that were separated by the line. These codes were entered into the data base as the line was digitized. The display functions that were written selected the specific line types from the data base, and then plotted them in a choice of four colors and a variety of line types which employed standard Calcomp calls. The initial design stage was chosen for its convenience and ease in meeting the original 305 grant's demand for an operational system within a modest budget. Little time and money were available for research and evaluation of existing systems elsewhere. The need for site-specific data precluded the use of crude grid cell systems that were being utilized for other applications.[6] At that time, polygon-based concepts were not well-established at a broad regional scale and were largely confined to line printer routines, such as SYMAP.[7] The decision to use the line segment format was based upon a desire to replicate and update lines on a 1:24,000 quadrangle. The basic procedures utilized rather straightforward digitizing and display operations that appeared to meet the initial objectives that had been outlined in the proposal.

DEVELOPMENTS AT THE UNIVERSITY OF SOUTH CAROLINA

In Late 1974 when the Coastal Zone Mapping Program outgrew the resources of the Charleston County Planning Office, the decision was made to move it to the University of South Carolina Computer Services Division in Columbia. The change, which involved the relocation of both the digitizing and plotting equipment to the main installation, resulted in a greatly expanded computer capacity (IBM 370/168). The move also offered the opportunity to develop on-line graphical digitizing and editing functions.

The role of major state universities in the development of state-sponsored automated systems is a provocative one. It is difficult for a state-supported institution to ignore the needs of

[5]"Coastal Zone Management Program Development Grants," Federal Register, V. 38, No. 220.

[6]Numerous grid cell systems were in operation in different states. See for example: George Orning and Les Maki, Land Management Information in Northwest Minnesota (Minneapolis, Minnesota: Minnesota Land Management Information System, 1972); Ronald L. Shelton and Earnest E. Hardy, "The New York State Land Use and Natural Resources Inventory," Proceedings of the Seventh International Symposium on Remote Sensing of Environment, (Ann Arbor, Michigan, 1972), pp. 1571-1575; and David J. Cowen Development and Applications of the South Carolina Computerized Land Use Information System, (Columbia, S.C.: South Carolina Land Resources Conservation Commission, 1976).
[7]Laboratory for Computer Graphics and Spatial Analyses, SYMAP: Users Reference Manual, (Cambridge, Massachusetts: Laboratory for Computer Graphics, 1975).

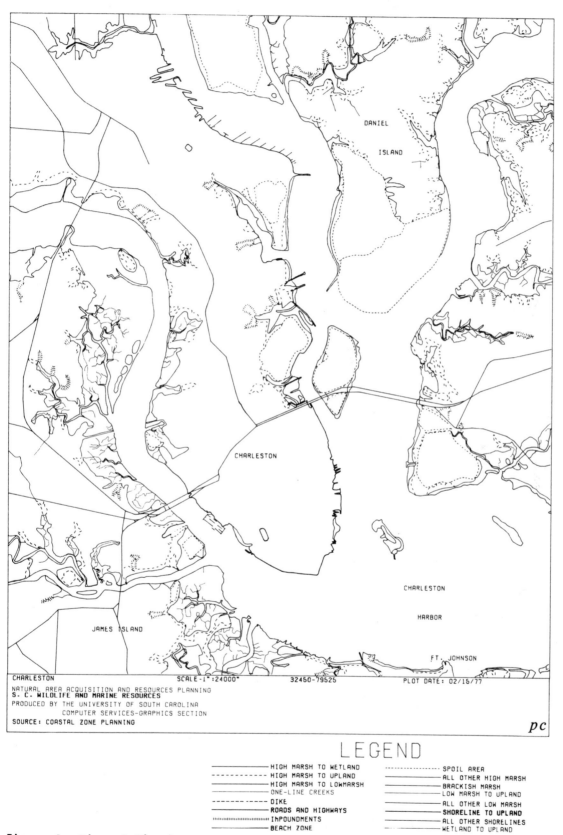

Figure 1. Plot of Charleston, S.C. quadrangle. This plot demonstrates the ability to display boundary line segments as different line types. Original plot in four colors.

the various state agencies, especially when they are physically adjacent to each other in the same city. More important, public service activities have recently assumed an increasingly significant role in such universities. Consequently, the research and development activities of their faculty and staff also reflect this interest. In fact, many significant developments in the field of automated cartography and geographical data handling in general have emerged from just such institutional settings.

In addition to maintaining an excellent hardware system, the University of South Carolina Computer Services Division made a conscious commitment to the development of a graphics system that would serve both the public and academic communities. The significant capital investments and expensive research and development activities were shared by the two sectors to their mutual benefit.

The development of the Coastal Mapping Program at the University was shaped by the constraints of already existing hardware and software systems. The University was basically a large IBM shop, oriented to batch processing. There was considerable disk storage available and IMS was available for data base management. APL, which was supported on the system, offered an interactive language that had graphics capabilities. Using these resources the graphics staff created a set of procedures for the input of digitized data, on-line graphical display and editing of the data, selective data retrieval and preparation of data for plotting. In fact, the on-line editing functions developed by the staff represent one of the only such APL-based systems in existence. Significant improvements in data entry, editing, handling, and plotting resulted from the new software developments. Since many of the functions could be quickly displayed on a Tektronix storage tube it was possible to demonstrate the system at remote locations via telephone lines. A concerted effort was made to publicize the system using a demonstration of these interactive capabilities. Subsequently, the system has been enhanced through the addition of an IBM mass storage device, a Princeton graphics tube, a Gould electrostatic plotter, a backlighted digitizing table and, most recently, a Kongsberg 5000 flatbed plotter that has scribing capabilities. Software is now maintained through IBM's VSPC system. The total system represents a highly sophisticated and efficient on-line graphics system with tremendous data storage and management capabilities.

Although the basic objective of the system has continued to be the production of the updated 7 1/2-minute quadrangles (Figure 1), as other data emerged new refinements were necessitated. For example, archaeological and historical sites required the handling of point data. Experiments with the placement of symbols on the map were conducted to improve the recognition of the different data categories (Figure 2). By January of 1977, sixty of the sixty-eight quadrangles for which wetlands overlays had been produced were digitized and fifty-two had been plotted. These plotted products served as an important inventory and display mechanism during the legislative debate concerning South Carolina's own Coastal Zone Act. The graphical output dramatically demonstrated the rapid changes that had been occurring in the coastal zone. More significantly, the interactive system proved that automated methods could be efficiently utilized to handle geographically related data.

IMPACT OF THE SOUTH CAROLINA COASTAL ZONE MANAGEMENT ACT

The debate concerning the South Carolina Coastal Zone

Management Act culminated with its passage in the spring of 1977. Although it would be difficult to assess the direct impact of the coastal mapping program on the passage of the Act, there is little doubt that the requirements of the Act indicate considerable faith in the ability to inventory and model data relating to the coastal zone. Specifically, the Act charges the Coastal Council with:

(1) undertaking the related programs necessary to develop and recommend a comprehensive management plan;

(2) examining, modifying, improving or denying applications for permits for activities;

(3) Managing estuaries and marine sanctuaries and regulating all activities therein;

(4) establishing, controlling and administering pipeline corridors and location of pipelines used for the transmission of any form in a critical area;

(5) Directing and coordinating the beach and coastal shore erosion control activities.[8]

As part of the charge to develop the coastal management program, the following items were specified:

(1) identify present land uses and coastal resources;

(2) evaluate these resources in terms of their quality and quantity and capability for use both now and in the future;

(3) determine the present and potential use and the present and potential conflicts in use of coastal resources;

(4) inventory and designate areas of critical state concern in the coastal zone, such as port areas, significant natural and environmental, industrial and recreational areas.[9]

These functions imposed an immediate responsibility on the Coastal Council to develop certain analytical tools. An evaluation of the existing data base and software capabilities indicated that, although they were extremely efficient for graphical data entry editing, retrieval and display, they had not been designed to handle the type of analytical tasks outlined by the new law. The basic problem with the original system related to the assumption that a map consists of a series of lines rather than a set of areas. It must be remembered, however, that the original objective had been the development of a cartographic system. These new requirements called for an entire geographical information system in which display is merely one component.

RE-EVALUATION

In the summer of 1977, a serious research effort began to conceptualize and specify the requirements of a more adequate software system. Unlike the original design stage, this effort has been much more systematic. Of course, a number of changes have occurred since 1974. From the viewpoint of the Coastal Mapping Program, the most significant of these have been: (1) the evolution of a serious body of literature regarding computer cartography and geographical information systems; (2) a significant improvement in polygon-based geographical data handling procedures; (3) considerable experience regarding the capabilities and limitations of automated methods by both users and designers; and (4) the involvement of geographers

[8]"Coastal Tidelands and Wetlands," *Code of Laws of South Carolina 1976* (Supplement) Vol. 16, Chapter 39, Section 48-39-50, p. 431.
[9]Ibid, Section 48-39-80, p. 43.

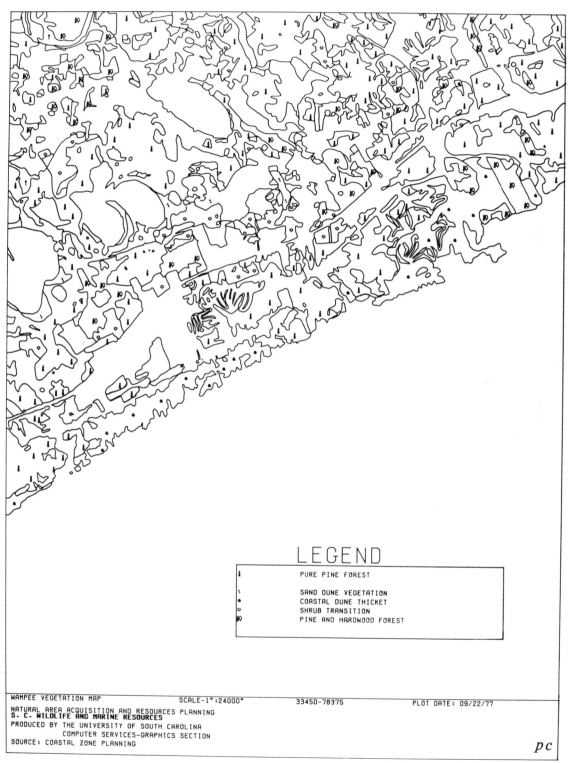

LEGEND

ᛁ	PURE PINE FOREST
\	SAND DUNE VEGETATION
•	COASTAL DUNE THICKET
o	SHRUB TRANSITION
ꓕ0	PINE AND HARDWOOD FOREST

WAMPEE VEGETATION MAP   SCALE-1":24000"   33450-78375   PLOT DATE: 09/22/77

NATURAL AREA ACQUISITION AND RESOURCES PLANNING
S. C. WILDLIFE AND MARINE RESOURCES
PRODUCED BY THE UNIVERSITY OF SOUTH CAROLINA
         COMPUTER SERVICES-GRAPHICS SECTION
SOURCE: COASTAL ZONE PLANNING

pc

Figure 2. **Plot of vegetation for Wampee, S.C. quadrangle. This plot demonstrates the use of symbols to help identify different areas. Original plot in four colors.**

and cartographers in the conception and design of the new system. Each of these factors has greatly improved the changes for designing and constructing a successful system.

Thus far, the evolution of the new cartographic and analytical software has emerged through the concept and design stages. After a careful review of the relevant literature, considerable discussion with system users, administrators and designers, and extensive interviews and consultations with other groups with similar needs, the basic system objectives were outlined. These goals suggest the need for a set of polygon-based procedures that will allow for: (1) flexible entry of point, line and area data, (2) ability to handle various coordinate systems, (3) ability to analyze, manipulate and display polygons, and (4) ability to interrelate and model various data elements. The underlying assumption behind this search process has been that the state-of-the-art has reached the point where such a polygon-based system should now be feasible.

The development of specific components of an improved system has been based on a positive assessment of the existing system components which would provide the basis for any changes. These components included an excellent hardware configuration, an experienced programming staff, and an excellent data entry, editing, storage and display system. These assumptions mandated that any new software would be installed on the IBM mainframe and would have to be available in source code. With much of the recent developmental activity in the area of geographical information systems being geared to dedicated mini-computers, these constraints proved to be quite restrictive.

Pressures from various sources dictated that the step from the concept and design stages to the construction and operationalization stages occur more rapidly than would have been desired. From the federal perspective, South Carolina was in the final year of eligibility for 305 development funds. Future resources would be dependent upon the implementation and administration of the Section 306 management plan. From the state perspective, there was already a backlog of permit applications awaiting systematic evaluation. Therefore, although there was confidence that the staff of the University could develop its own software, the timetable precluded an in-house effort and a search outside the University began. The need for system support, training and rapid installation eliminated the inclusion of software available from public sources or other universities from the search. This decision was particularly regrettable in light of the excellent developments occurring within several federal agencies, foreign countries, and universities. The evolutionary process has been completed through the search stage and is presently commencing the testing and evaluation phase. Although the constraints imposed by the specifications allow for very few commercially available software systems, there remains considerable optimism that new software can be installed and tested this summer. Unfortunately, geographical software developments still lag considerably behind those on the hardware side and, as Dueker suggests, methods for evaluation of such software are not well-established.[10]

ANALYSIS OF SYSTEM DEVELOPMENT PROCESS

The idea that complex ecological systems can be understood only when sophisticated electronic data-processing equipment is employed is an especially appealing one to those involved with computer technology. However, the success of such applications has been mixed at best. As the American Society of Planning Officials cites, "most states apparently are not in a position to use automated data systems efficiently at this time. Base data are too variable and in some cases incomplete. Interrelationships between factors are not yet sufficiently understood and defined and time and money are short."[11] Therefore, although Turner, et al. suggest, "it is difficult to locate an influential planner who publicly deviates from the view that computerized information systems are about to rationalize if not revolutionize the planning process,"[12] skepticism still abounds. One author goes so far as to assert that "most practicing planners have found the magic maker to be a more useful tool."[13] Thus, in evaluating the South Carolina Coastal Mapping Program, one must recognize the imprecise nature of the science which has only recently developed a group of theoreticians and practitioners of sufficient number to produce a meaningful dialogue on the basic concepts and applications.

IDEAL MODEL

A fair assessment of the program should be conducted only through a comparison with an idealized model of how such a beast should evolve. From the works of people such as Tomlinson, Marble, Calkins, Kennedy, Guinn, Meyers, and Dueker,[14] it is now possible to discern an idealized developmental model for such an undertaking.

The initial stage in the evolution of any system should involve a statement of need on the part of a group or users. Typically, the initiation of a natural resource information system has emanated as a response to a specific planning function. Dueker suggests that there are six such functions: (1) policy planning, (2) program planning, (3) land inventory, (4) impact analysis, (5) land capability analysis, and (6) regulatory activities.[15] Intuitively, it is believed by many that automation will help ac-

[10]Kenneth Dueker, *Geographic Data Encoding Issues* (Iowa City, Iowa: The University of Iowa Institute of Urban and Regional Research, 1975).

[11]American Society of Planning Officials, *Information/Data Handling Requirements for Selected State Resource Management Program* (Washington, D.C.: U.S. Department of the Interior, 1975).

[12]W. Turner, E. Hollander and J. Getzels, "Stonewalled: Barriers to Adoption of Computer Oriented Land Use Information Systems," *Issue Papers (Technical Report E,* (Washington, D.C.: U.S. Department of the Interior, 1975).

[13]Charles Kindleburger, "Urban Planning Information Systems: Name One That Works," Paper presented at American Institute of Planners 57th Annual Conference, 1974.

[14]See for example: International Geographical Union Commission on Geographical Data Sensing and Processing, *Second Interim Report on Digital Spatial Data Handling in the U.S. Geological Survey* (Reston, Virginia: U.S. Geological Survey, 1976); Michael Kennedy and Charles Meyers, *Spatial Information System: An Introduction* (Louisville, Ky.: Urban Studies Center, 1975); Charles Guinn and Michael Kennedy, *Avoiding System Failure: Approaches to Integrity and Utility* (Louisville, Ky.: Urban Studies Center, 1975), and Kenneth Dueker and Richard Talcott, *State Land Use Planning Process Issues: Geographic Information System Implications* (Iowa City, Iowa: University of Iowa Institute of Urban and Regional Research, 1975).

[15]Kenneth Dueker, *Geographic Data Encoding Issues* (Iowa City, Iowa: University of Iowa Institute of Urban and Regional Research, 1975), p. 8.

complish any one or all of these functions better, quicker, and/or cheaper than manual procedures. Once the user group has begun to outline a conception of the system, it usually produces a broadly defined, grandiose goal, such as "to provide better information for wise decision making." Armed with this type of vague goal statement, the users attempt to attract a sponsoring group. Considering the number of federal, state and local programs requiring analyses of geographical data, sponsors willing to support an innovative approach to data handling have not been too difficult to find in recent years.

Once the financial commitment has been assured, the users, in conjunction with designers, data collectors and administrators, begin to develop specific objectives for the system. These objectives usually address one or more of the six tasks outlined by Dueker: (1) measurement, association, and display; (2) record keeping and monitoring; (3) locational analysis; (4) diffusion studies; (5) spatial interaction; and (6) trend projection.[16] In order to accomplish any of these tasks, Marble and Calkins suggest that several components must be designed. In fact, they maintain that separate sub-systems must be developed to handle problems relating to: (1) management, (2) data acquisition, (3) data input, (4) retrieval and analysis, (5) information output, and (6) information use.[17] At this stage, precise determination of the various components and parameters of the system must also be defined. After a thorough search of existing methodological approaches has been conducted, several alternatives should be selected for detailed evaluation in terms of their individual benefits and costs. It is only after a detailed benchmark test has been conducted that the system can be prepared. During a pilot study, the users, sponsors, designers, data collectors and administrators should work closely together in order to establish the proper feedback channels which will promote each individual's understanding of his role in the system, and empathy for the problems encountered by the others.

The idealized system, of course, would never be operationalized until each of the sub-systems has been tested, and it has been demonstrated that they can perform together as an efficient system. In order for this to occur, all of the appropriate training sessions must have been conducted; documentation of each aspect must be available in a looseleaf notebook; a well-managed public relations plan must have been implemented, both to inform users of the benefits of the system, and to ensure that the sponsor can demonstrate the value of the new systems; and marketing and distribution policies that will maximize the utilization of the system over the first five years must be well-formulated. Additionally, data collectors in various agencies must have been convinced that only through the collection of accurate, timely and relevant data will wise decisions be made. The public, too, must be assured that only through the use of the modern system can we possibly expect to achieve a perfect balance between economic development and environmental protection. Finally, the designers and implementers must have been allotted at least ten years had several million dollars to bring the system into full production.[18]

---

[16]Ibid., p. 18.

[17]International Geographical Inion, Commission on Geographical Data Sensing and Processing *Second Interim Report on Digital Spatial Data Handling in the U.S. Geological Survey* (Reston, Va.: U.S. Geological Survey, 1976), p. 20.

[18]These are estimates of time and cost figures suggested by Michael Kennedy and Charles Meyers, *Spatial Information Systems: An Introduction* (Louisville, Ky.: University of Louisville Urban Studies Center, 1975), p. 64.

EVALUATION OF THE SOUTH CAROLINA EXPERIENCE

Such idealized models serve as valuable reminders of the enormous gaps that exist between expectations and reality. As suggested by the description of the evolution of the South Carolina Coastal Mapping Program, severe constraints, in terms of deadlines, financial and personnel resources, understanding and communication of needs, and knowledge of concepts and institutional arrangements, resulted in a developmental process that deviated significantly from this model. However, the dominant question that must be addressed before evaluating a program is "how close to the model can one realistically expect to come?". When compared to the model, the returns for the investment of approximately $200,-000 and a four-year effort must be considered significant. In addition to the tangible products represented by the updated 7 1/2-minute quadrangles, there have been numerous and extremely beneficial results from the process itself. Of necessity, many of these returns are difficult to measure in the short run. One must consider that, while the program commenced practically in a vacuum, it has made significant progress toward the present state-of-the-art. The framework for an excellent construction stage for an upgraded system has been established. Although all of the objectives for the system have not been formulated (and can be expected to continue to change periodically), there now exists a realistic set of specifications that are in line with expectations.

A serious evaluation of the causes for the disparity between the ideal model and reality is a luxury available only in retrospect. However, such an assessment should be valued as the best source of education, both for avoiding future mistakes and for pinpointing areas which need improvement. By far the greatest obstacle to the successful implementation of the ideal model was created by the unrealistic demands placed on the system by the Office of Coastal Zone Management. The 305 development grant's emphasis on the use of existing data and its three-year timetable have greatly penalized those states that were forced to begin with insufficient data bases and have attempted to implement innovative data handling procedures. Conversely, those states that undertook modest, one-time inventory efforts and met deadlines by employing conventional methods have been rewarded. It must be assumed that, in the long run, the latter will become less effective mechanisms for assessing management decisions. An additional set of obstacles can be identified which pertain specifically to the South Carolina experience but probably are symptomatic of any such undertaking. Among the most significant of these are:

(1) unclear specification of system needs by the user;

(2) limited or poor understanding of geographical data concepts on the part of designers;

(3) overselling of system capabilities;

(4) unclear demarcation between research and development, and production;

(5) physical distance (of 120 miles) between users and production staff;

(6) unclear definition of responsibilities;

(7) difficult verbal communications between users and production staff;

(8) poor hardware performance;

(9) new hardware components;

(10) evolution of geographical data base concepts.

Many of these obstacles were to be expected and could be resolved only in what Dueker and Talcott consider to be an incremental approach.[19] In fact, one theory in public administration suggests that this type of evolution is characteristic of any learning experience that necessarily involves a great deal of "muddling through."[20]

In an effort to alleviate some of the organizational obstacles relating to the system, a series of management changes have been implemented. Geographers and cartographers who are familiar with map design, spatial relationships and applications, have been installed in key positions to aid in supervision of map preparation, digitizing operations and communications. Research and development staff have been separated from production operations. Responsibilities have been more clearly defined. For example, contractual arrangements, including specification of inputs and outputs and scheduling, are being handled by the Office of Geographical Statistics of the Division of Research and Statistical Services. That office, which includes the state cartographer and state geographer, also provides the staff for the state's Mapping Advisory Committee. As such, it can provide an essential link between the University Computer Services Division and the Coastal Zone Council. By serving as the liaison group, it can reduce ambiguities that arise, while also giving the staff of the computer center more specific guidelines and priorities. User groups are sheltered from direct communication with computer scientists. Promotion of the system is now being carried out in a more conservative fashion. Although considerable uncertainty still exists regarding the implementation of new software components, decisions are now being made in a rational manner with much greater confidence that the system will be able to respond to future needs. The partnership between the University and state government can be expected to strengthen and expand as new demands for such technical support arise.

BIBLIOGRAPHY

AMERICAN SOCIETY OF PLANNING OFFICIALS. *Information/Data Handling Requirements for Selected State Resource Management Programs*. Washington, D.C.: U.S. Department of the Interior, 1975.

COWEN, DAVID. *Development and Applications of the South Carolina Computerized Land Use Information System*. Columbia, South Carolina: South Carolina Land Resources Conservation Commission, 1976.

DUEKER, KENNETH. *Geographic Data Encoding Issues*. Iowa City, Iowa: University of Iowa Institute of Urban and Regional Research, 1975.

———, and TALCOTT, RICHARD. "An Incremental Approach to the Design of a Geographic Information System," *1977 URISA Proceedings* III (1977): 298-305.

GUINN, CHARLES and KENNEDY, MICHAEL. *Avoiding System Failure: Approaches to Integrity and Utility*. Louisville, Kentucky: Urban Studies Center, 1975.

INTERNATIONAL GEOGRAPHICAL UNION COMMISSION ON GEOGRAPHICAL DATA SENSING AND PROCESSING. Second Interim Report on Digital Spatial Data Handling in the U.S. Geological Surveys. Reston, Virginia: U.S. Geological Survey, 1976.

KENNEDY, MICHAEL and MEYERS, CHARLES. *Spatial Information Systems: An Introduction*. Louisville, Kentucky: Urban Studies Center, 1975.

KINDLEBURGER, CHARLES. "Urban Planning Information Systems: Name One That Works," Paper presented at the American Institute of Planners 57th Annual Conference, 1974.

LABORATORY FOR COMPUTER GRAPHICS. *SYMAP: Users Reference Manual*. Cambridge, Mass.: Laboratory for Computer Graphics, Harvard University, 1975.

LINDBLOM, CHARLES E. "The Science of Muddling Through," *The Public Administration* 19 (1959): 79-88.

ORNING, GEORGE and MAKI, LES. *Land Management Information in Northwest Minnesota*. Minneapolis, Minnesota: Minnesota Land Management Information System, 1972.

SHELTON, RONALD L. and HARDY, ERNEST E. "The New York State Land Use and Natural Resources Inventory," *Proceedings of the Seventh International Symposium on Remote Sensing on Environment*. Ann Arbor, Michigan, 1972.

SOUTH CAROLINA COASTAL ZONE PLANNING AND MANAGEMENT COUNCIL. "An Application for a Program Development Grant to the Office of Coastal Environment." March, 1974.

TURNER, W.; HOLLANDER, E., and GETZELS, J. "Stonewalled: Barriers to Adoption of Computer Oriented Land Use Information System." *Issue Papers (Technical Peport E.)* Washington, D.C.: U.S. Department of the Interior, 1974.

---

[19]Kenneth Dueker and Richard Talcott, "An Incremental Approach to the Design of a Geographic Information System," in *1977 URISA Proceedings*, pp. 298-305.

[20]Charles E. Lindblom, "The Science of Muddling Through," *The Public Administration Review*, Vol. 19, (1959), pp. 79-88.

Introductory Computer Mapping Instruction at the University of South Carolina

by David J. Cowen

INTRODUCTION

The production of maps by computer-assisted methods has become a widely accepted practice in government, as well as in private and educational environments. Computer-produced maps are now found in numerous planning reports and research publications. Although the history of computer cartography may be considered to span less than twenty years, the field seems to abound with new, revised or refined software. For example, the IGU Commission on Geographical Data Sensing and Processing has discovered more than 170 different computer mapping software packages.[1] Concurrently, the field has been developing a sizeable pool of researchers and a body of serious scientific literature. In fact, it has been conjectured that more than 3,000 articles dealing with computer mapping have been prepared.[2]

Although there has been considerable debate concerning the definitions of the field of computer-assisted cartography, it still appears to range anywhere from the production of line printer maps in local planning agencies to extensive digitizing, scribing or photo-head plotting by Federal agencies. Despite some semantic differences between computer cartography, automated cartography, computer-aided cartography and computer mapping, the fact remains that an extensive variety of organizations have found the production of maps with some assistance by a computer to be beneficial. Rhind lists eleven different reasons for the rapid growth of the field.[3] These encompass the typical concerns for speed, accuracy, and cost, as well as certain institutional considerations. It is interesting to note that both cost and institutional considerations also appear among Rhind's list of disadvantages with the production of computer-generated maps.

As the field has expanded and come out of the "research and development closet," an increased demand for educational and training programs has been a logical consequence. This demand has been partially met by the development of several formal university courses that deal with the subject. Although Dahlberg found only eleven such courses listed in the college catalogues in the United States in 1975-76, this number represents only those with the term "computer" in the title.[4] Therefore, his list systematically excludes a large number of advanced cartography courses that include major sections on the use of automated methods. The purpose of this paper is to report on the evolution, objectives, method, and experiences relating to the development and presentation of a computer mapping course at the University of South Carolina. Hopefully, the experiences relating to this particular course will be of value to others who are faced with similar developmental problems.

EVOLUTION

The development of a course in computer mapping at the University of South Carolina emerged directly from the research and service activities of a few members of the Geography Department.[5] In 1970 the line printer represented the only graphical output device available at the University of South Carolina, and computer capacity was restricted to an IBM 7090 with 32K storage. Thus constrained, Morton Scripter's CMAP program formed the only functional computer-mapping software available at that time.[6] That program provided a basis for several graphical presentations that were produced for local and state agencies. It was demonstrated that even those simple line printer maps, when used in conjunction with an overlay, could produce effective presentations of statistical data for various types of publications.[7]

As the computer capacity was upgraded, several modifications were made to CMAP that facilitated the access of statistical and outline data from various modes, while also providing internal debugging procedures. At approximately the same time, the GRIDS[8] program was acquired from the Census Bureau and the Harvard programs (SYMAP,[9] CALFORM,[10] GRID,[11] SYMVU[12]) were acquired with the aid of a Ford ve-

[5]The development of the course and many applications have been accomplished in conjunction with Paul E. Lovingood, Jr.

[6]Morton W. Scripter, "Choropleth Maps on Small Digital Computers," *Proceedings of the Association of American Geographers* I (1969): 133-136.

[7]Paul E. Lovingood, Jr., "Computer Produced Choropleth Maps with Overlays: An Aid in Urban and Regional Analysis," *The Greek Review of Social Research* 24, (1975): 311-319.

[8]Mathew A. Jaro, *GRIDS — A Computer Mapping System* (Washington, D.C.: U.S. Department of Commerce, Bureau of the Census, 1972).

[9]Laboratory for Computer Graphics and Spatial Analysis, *SYMAP: Users Reference Manual* (Cambridge: Harvard University Laboratory for Computer Graphics and Spatial Analysis, 1975).

[10]Laboratory for Computer Graphics and Spatial Analysis, *CALFORM Manual* (Cambridge: Harvard University Laboratory for Computer Graphics and Spatial Analysis, 1972).

[11]Laboratory for Computer Graphics and Spatial Analysis, *GRID Manual Version 3* (Cambridge: Harvard University Laboratory for Computer Graphics and Spatial Analysis, 1971).

[12]Laboratory for Computer Graphics and Spatial Analysis, *SYMVU Users-operators Reference Manual* (Cambridge: Harvard University Laboratory for Computer Graphics and Spatial Analysis, 1971).

[1]Kurt Brassel, *A Survey of Cartographic Display Software*, Report R-77/3 (Buffalo, N.Y.: Geographic Information Systems Laboratory, 1977).

[2]K. H. Meine, personal communication referred to in David Rhind, "Computer-aided Cartography," *Transactions, Institute of British Geographers, New Series* 2, (1977): 71-97.

[3]David Rhind, "Computer-aided Cartography," *Transactions, Institute of British Geographers, New Series* 2, (1977): 71-97.

[4]Richard E. Dahlberg, "Cartographic Education in U.S. Colleges and Universities," *The American Cartographer* 4 (1977): 145-156.

ture grant. Further experience was gained through additional local public service projects. As students and faculty members throughout the University became aware of the existence of these mapping programs, the need for a more systematic training method became obvious. Therefore, from its inception, the course has concentrated on the training and application aspects of computer mapping rather than on the theoretical or programming concerns. The primary objective has been to provide a practical working understanding of several of the most commonly used computer mapping packages. Instructional emphasis has concentrated on the preparation of the necessary inputs which would subsequently yield the final map products.

COURSE PHILOSOPHY

The course is basically designed to meet the minimum needs of a producer of computer maps. Such producers may emerge from a variety of academic fields with any degree of familiarity with computer technology. In order to meet this need, the course has no prerequisites and is available to both undergraduate and graduate students. The course meets once a week during a thirteen-week semester and is now offered every semester. It has been offered both in late afternoon or early evening to accommodate part-time students who hold full-time jobs during the day. Enrollment has increased markedly and now is restricted to thirty students per section. Although enrollment in the course has come from various parts of the campus, the disciplines of geography, geology, marine science, computer science, sociology, community psychology, public health, engineering, and criminal justice represent the most sizeable groups.

COMPUTING ENVIRONMENT

The University presently operates an IBM 370/168 batch-oriented computer center. With tape, disk and IBM mass storage devices available, problems related to storage constraints have been eliminated. Most of the work for the course is handled in the Social and Behavioral Sciences Laboratory, which is a college-supported facility. The lab operates a DATA 100 card reader and printer, as well as a GOULD 12-inch electrostatic plotter. That plotter, which is CALCOMP compatible, now provides rapid turn-around for plot routines that previously required more than twenty-four hours on the drum plotter at the main computer center. Although the University does not support a fully interactive system such as TSO, it is now running IBM's VSPC which enables the user to submit and retrieve batch jobs through CRT devices. The lab is geared to the accommodation of mapping software, such as SYMAP, that is often printed at eight lines per inch. Various types of consultation are available from student and staff personnel in the lab. In addition, students have access to a GRAPH-PEN sonic digitizer that is attached to a keypunch in the Geography Department.

COURSE CONTENT

Although the content of the course varies somewhat every semester, it has continued to be oriented toward the use of common software packages. Although the original course consisted of CMAP, GRIDS, SYMAP and SYMVU, other routines, such as Tobler's RGRID,[13] a DIME[14] file plotting program and CALFORM have frequently been included. With the installation of the electrostatic plotter last summer, the time devoted to CALFORM and SYMVU has increased in proportion to the other routines. Unfortunately, in order to accommodate this, the GRIDS and DIME routines have had to be eliminated.

The basic method of presentation emphasizes the preparation of required data, map and control card inputs. To facilitate the understanding of the essential base-map inputs, each student is assigned a simple geometric figure that is divided into five or six sub-areas. As each routine is presented, the student is required to produce a base map for his figure. For CMAP this means that each student prepares a set of scan lines, for SYMAP A-Conformolines, and for CALFORM the points and polygons packages. The purpose of the geometric figure is to provide the experience of base-map preparation without the drudgery of tedious bookkeeping and card preparation. Each student is also required to gather and prepare at least a three-variable data set relating to the forty-six counties of South Carolina. Subsequently, this data set is employed to access South Carolina base maps that are stored on disk for the CMAP, CALFORM, and SYMAP routines. For each mapping program, the student progresses from the geometric figure to the production of a South Carolina map.

Typically, the core of the course has revolved around the various applications of SYMAP. Students learn the basic input requirements for choropleth, proximal and contour maps by working with their geometric figure. They also become familiar with C-otolegends, C-legends and other options. After they have mastered the basic concepts, they are introduced to FLEXIN subroutines and methods for accessing data stored in nonstandard modes and formats. The course has now progressed to the stage where students are required to use SPSS[15] to access some of the more than 200 variables in the County-City Data Book[16] at the county level for any of the forty-eight contiguous states. These variables are passed into SYMAP through a FLEXIN subroutine and linked to the corresponding A-Conformolines package that has been created through POLYVRT[17] and is stored on tape. SYMVU has been modified to enable it to be run as a subsequent step of a SYMAP job. At this stage of the course, the student has an enormous potential for producing an almost infinite variety of maps just through the preparation of a few simple control cards.

The preparation of complex base maps has been facilitated by the development of a special routine that prepares the necessary packages for SYMAP or CALFORM from digitized data. This routine, TODDTOS (Translation of Digitized Data To Other Systems), provides a simple procedure for the single digitizing of node and chain files on the GRAPH-PEN machine. The only required inputs are a node file, a chain file and a polygon file, each of which is in sequential order. Polygons are defined by the appropriate combination of nodes and chains

[13]Waldo Tobler, *Selected Computer Programs* (Ann Arbor: University of Michigan Department of Geography, 1970).

[14]U.S. Bureau of the Census, *The DIME Geocoding System*, Census Use Study Rep. 4. (Washington, D.C.: Department of Commerce, Bureau of the Census, 1970).

[15]Norman Nie, et al., *SPSS, Statistical Package for the Social Sciences*, 2nd Ed. (New York: McGraw-Hill Book Company, 1975).

[16]U.S. Bureau of the Census, *County and City Data Book 1972, A Statistical Abstract Supplement* (Washington, D.C.: U.S. Department of Commerce, Bureau of the Census, 1973) (magnetic tape).

[17]Laboratory for Computer Graphics and Spatial Analysis, *POLYVRT: User's Reference Manual* (Cambridge: Harvard University Laboratory for Computer Graphics and Spatial Analysis, 1974).

that have been numbered for identification purposes. Chains that have been digitized in the "wrong" direction are reversed simply by the use of negative numbers. Once the appropriate node and chain files have been prepared, a variety of polygons can be created or omitted. The system has proved to be of great benefit in solving the problems of "double digitized" common borders in SYMAP and the tedious polygon definitions used in CALFORM. With appropriate job control, it is possible to successively run TODDTOS, SYMAP and SYMVU in the same job submission. Furthermore, these inputs can be linked with data stored on an external file.

One special routine that has been developed for the course is the DIME-file mapping routine. The DIME routine utilizes the Census Bureau ADMATCH routines and a line printer plot program. Inputs consist of street numbers, names and types which are prepared in a standard format. Standardized output has been limited to a list of initial data, rejected records, accepted records with applicable geocodes for census tract, enumeration district and block number, as well as the estimated X,Y coordinates, in both inches and state plane coordinates. The final output is the line printer plot which can be registered to a census tract overlay. Although the routine is simple to use, it provides a useful geocoding exercise for the student. Part-time students who work in local government agencies have found this routine to be especially beneficial.

The Census Bureau's GRIDS program provides an example of geographic data processing that falls more closely in the realm of geographical information systems than computer mapping. By employing SPSS as a data retrieval and manipulation package as a first step, GRIDS, provides an excellent display program for grid cell data. The keyword format that GRIDS uses, as well as the internal MAPTRAN Language, provide the student with a powerful set of analytical and display functions in an English-language type of input method. The SPSS and GRIDS combination has been linked to a statewide Land Use information file (CLUIS)[18] to form a realistic classroom experience in geographical information handling and display.

EVALUATION

Students are evaluated on the bases of weekly assignments and the preparation of a final project. As with many educational experiences, evaluation is often the most arduous task. This chore is particularly burdensome in a course in which the students' experiences and backgrounds are so varied. The evaluation philosophy is based on the definition of a minimum acceptable performance. This performance level is established by the weekly homework assignments that a student is expected to successfully complete in order to receive a grade of C. Each weekly assignment consists of the actual production of at least one computer-produced map. These maps are placed in a computer output folder that is checked periodically. At the end of the semester, the folder should contain examples of the setups and the final output that the student can use for a future reference guide.

Grades higher than a C are awarded to students who elect to work on a final project. These projects, which are defined by the students with guidance from the instructor, provide an excellent vehicle for demonstrating the ability to apply computer mapping software to a practical experience. The projects have varied

greatly in complexity and subject matter, ranging from land use mapping to shipwrecks. Several computer science students have developed improved procedures and new algorithms. These have included a graduated circle routine, a new version of CMAP, a three-dimensional plotting procedure, and a procedure for producing scan lines from A-Conformolines. Several original base maps have been digitized using the TODDTOS procedure and some students have used the opportunity to experiment with POLYVRT or the DIME procedure. The projects are presented to the class during the University's final exam period. Most importantly, the projects have graphically demonstrated that even students who have never previously keypunched can acquire sufficient mastery of the basic concepts of computer-map production to implement their knowledge in a practical manner. Invariably, the presentation of final projects has been an extremely rewarding experience for both the students and the teachers.

Although the basic objectives of the course are usually attained by the majority of the students, there are still a number of problems relating to such a course. Perhaps the most difficult question relates to the trade-off between theory and application. In thirteen sessions, what is the correct mixture of conceptual understanding and cookbook presentation? Although we have stressed the latter, we do not yet have a firm basis for evaluating its merits or disadvantages. Several students have been able to put their practical knowledge to work in their other course work, research, or employment. Furthermore, the folders and final projects represent an excellent method of demonstrating skills to prospective employers. Nevertheless, it is not clear whether, in the long run, the students would have benefitted more from a better understanding of a wider breadth of materials and applications. To meet some of these needs, a new course in geographical information systems has been introduced. This course will deal with urban and regional planning applications of geographical data bases. As such, it will include sections on data collection, encoding, storage, retrieval, analyses and display. It should provide a useful sequel to the computer mapping student who is seriously interested in geographical data handling.

Another significant problem relates to course materials. At present, the students are given a set of handouts that provide a basis for their understanding of the weekly assignments. These handouts often include actual printouts with examples of setups and the resultant outputs. The purpose of the handouts is to minimize discussion of the technical matters, such as job control language. Subsequent system modifications often necessitate the revision of the handouts, thereby frequently making them complicated and difficult to understand. Over time, we hope that the documentation and examples of setups can be standardized and placed in a booklet that will be available from the campus bookstore. Such a manual should include examples of common errors and instructions on appropriate corrections. In the meantime, the present mimeographed materials will have to suffice.

Inevitably, such a course consumes a great amount of time on the part of the instructor. In addition to the daily need to provide assistance in debugging or further instruction, there is the continual desire to improve the existing systems. Not only are there new options available and further integration of existing procedures to be undertaken, but there is also the need to examine entirely new procedures. For example, we are presently evaluating whether to incorporate SURFACE II[19] into

[18]David J. Cowen, *Development and Applications of the South Carolina Computerized Land Use Information System (CLUIS)* (Columbia, S.C.: South Carolina Land Resources Conservation Commission, 1976).

[19]R. J. Sampson, *Surface II Graphics System* (Lawrence: Kansas Geological Survey, 1975).

the existing course. Responsibilities for all of these tasks are not easily delegated to others. Graduate assistants for such a course are difficult to find, and in a masters degree program, can rarely be expected to serve more than two semesters. Furthermore, computer science students typically have not had experience with any of these special programs and must actually sit through the course before they can be of much assistance. So far we have been unable to find a really satisfactory solution to the problem of maintaining adequate assistance with the course, and it continues to place an inordinate amount of responsibility on the instructor.

The computer mapping course is now an established part of the geography curriculum at the University of South Carolina. As the demand for analysis and display of geographical data continues to grow in response to the requirements of the numerous social and environmental programs at all levels of government, the course can similarly be expected to increase in popularity. New devices, such as electrostatic plotters, multispectral scanners, automated digitizers, and color display units, coupled with an increased wealth of both cartographic and statistical data in digital form, indicate that despite all of the advances which have already occurred, we are just entering a take-off stage in terms of the evolution of the field. Although it is difficult to predict the full impact of computer mapping on university curricula of the future, our experiences with this course suggest that there is a great need for this type of training.

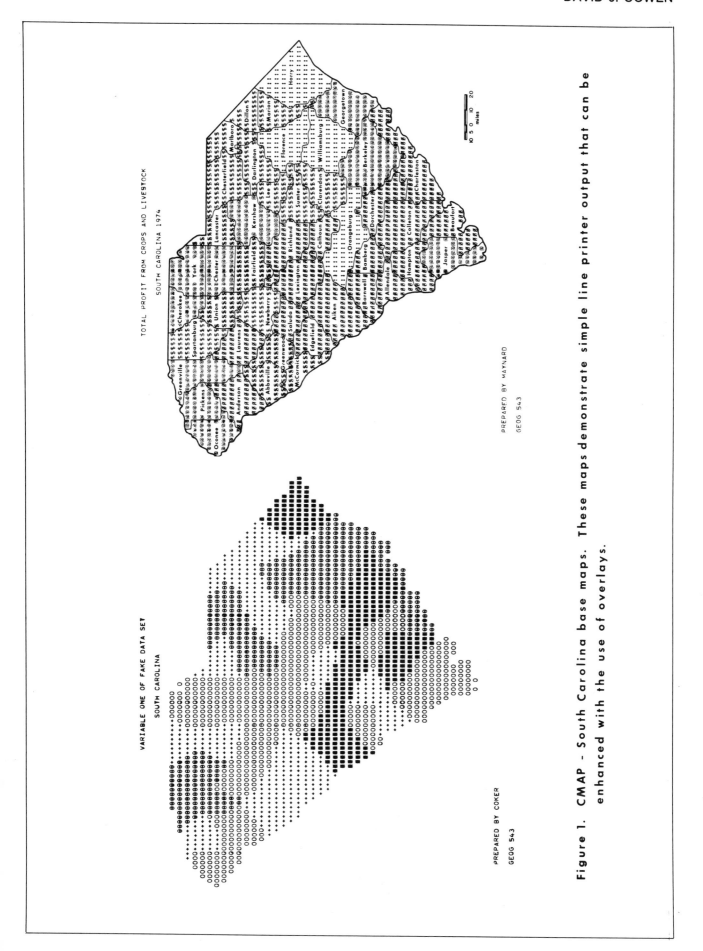

Figure 1. CMAP - South Carolina base maps. These maps demonstrate simple line printer output that can be enhanced with the use of overlays.

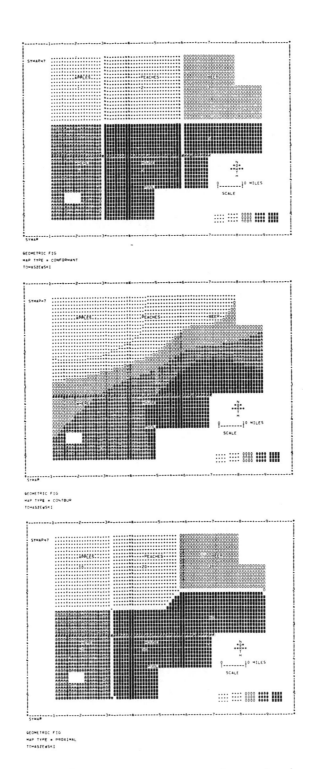

Figure 2. SYMAP conformant, contour, and proximal options using
geometric   figure with various C-Legends, C-Otolegends
and other options.

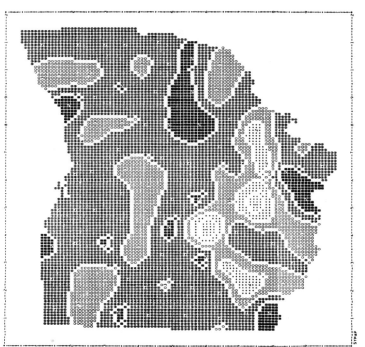

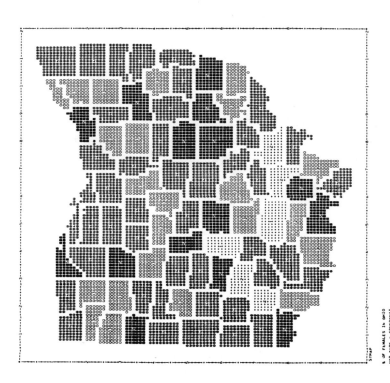

Figure 3. These maps demonstrate the ability of students to select any state and merge with data from the County - City Data Book via SPSS.

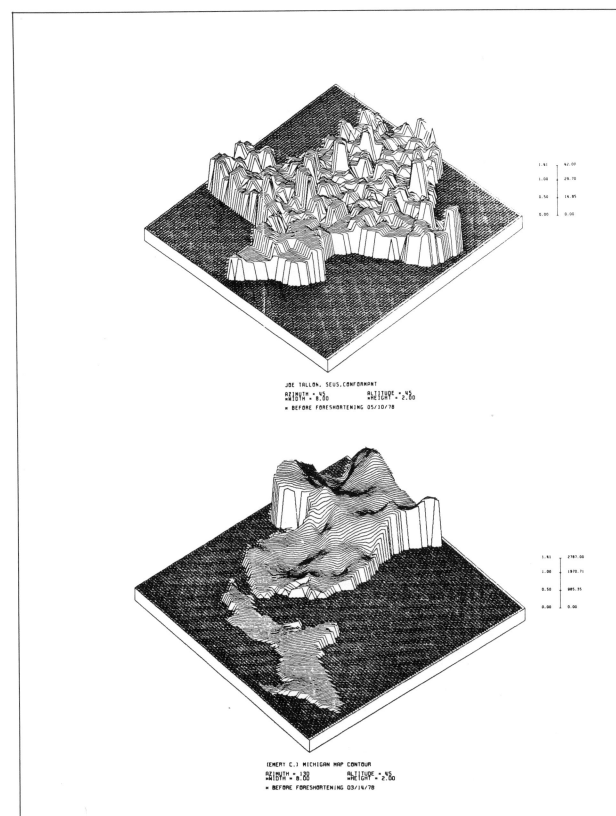

Figure 4. SYMVU plots on electrostatic plotter. These maps were generated in single runs that linked POLYVRT base, SPSS, SYMAP and SYMVU.

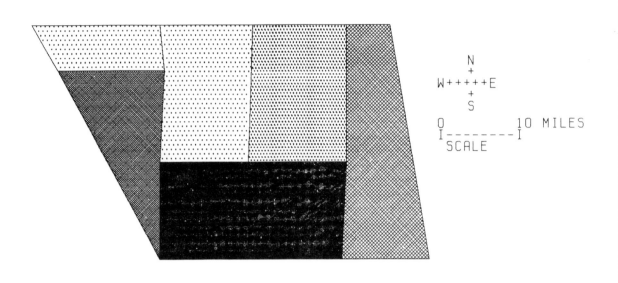

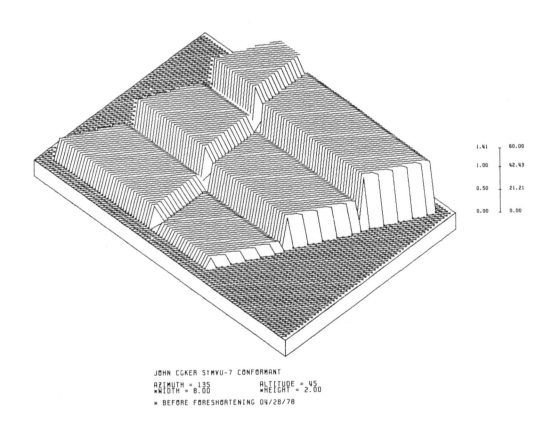

JOHN COKER SYMVU-7 CONFORMANT
AZIMUTH = 135   ALTITUDE = 45
*WIDTH = 8.00   *HEIGHT = 2.00

* BEFORE FORESHORTENING 04/28/78

Figure 5. CALFORM and SYMVU plots of geometric figures.

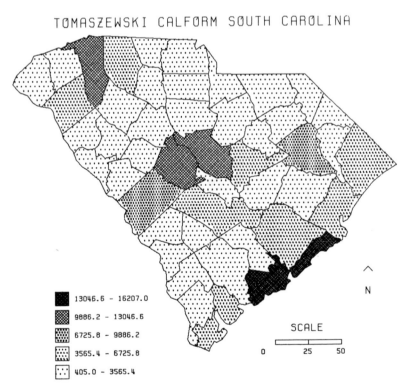

TOMASZEWSKI CALFORM SOUTH CAROLINA

13046.6 - 16207.0
9886.2 - 13046.6
6725.8 - 9886.2
3565.4 - 6725.8
405.0 - 3565.4

SCALE

0   25   50

N

Figure 6. **CALFORM** South Carolina maps on electrostatic plotter.

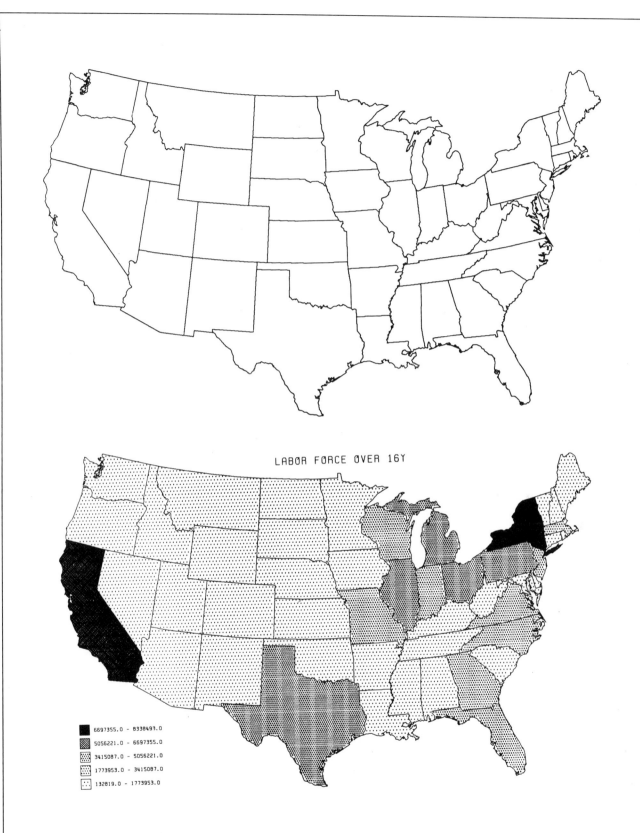

Figure 7   CALFORM U.S. maps on electrostatic plotter. Basemap
generated from POLYVRT data from County - City Data Book.

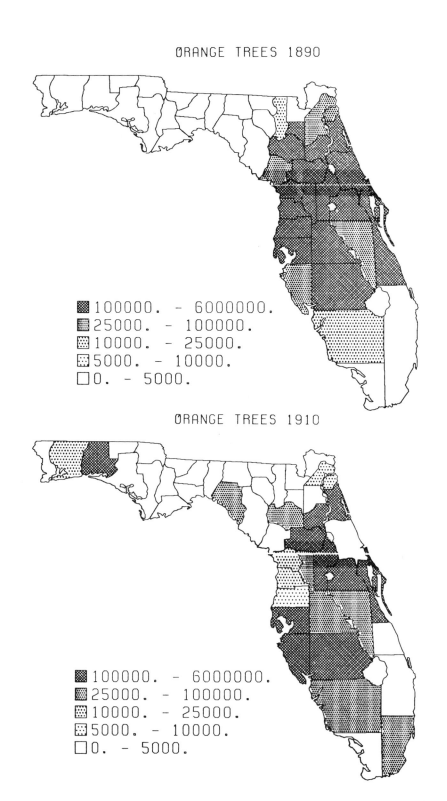

ØRANGE TREES 1890

■ 100000. - 6000000.
▨ 25000. - 100000.
▨ 10000. - 25000.
▨ 5000. - 10000.
□ 0. - 5000.

ØRANGE TREES 1910

■ 100000. - 6000000.
▨ 25000. - 100000.
▨ 10000. - 25000.
▨ 5000. - 10000.
□ 0. - 5000.

Figure 8. CALFORM maps on electrostatic plotter produced from digitized base map with different polygons defined in the **TODDTOS** procedure.

Difficulties in the Approach to the Mapping of the Environment in Cancer

by J.N.P. Davies, M.D.

Twenty-one years ago this month, my brother and I, working with the data of the late Dr. Gillian Jacobs, drew the first map of the distribution of that cancer which has come to be called the Burkitt lymphoma, and just 20 years ago, we recorded our findings.[1] Fundamental to our studies was a map made by E.S. Munger of the University of Chicago, who had been studying the relational patterns of Kampala in Uganda and had studied the homes from whence patients came to attend the clinics of Mulago Hospital, where eight years previously we had set up a Cancer Registry.[2] Thirty-six percent of those attending Mulago Hospital came from the mountainous southwest of Uganda; this figure coincided precisely with the overall figures of the cancer patients registered both for adults and for children, with the exception of the jaw lymphomas in childhood — none came from that area.[3]

Unfortunately, all the first cases of this lymphoma that we mapped came from the northern province of Uganda. Examination showed that by mapping the lymphomas in this way, we had mapped the radioactive areas of northern Uganda, a mischance that wasted two years of our work during which time others had advanced our knowledge. Such an experience was not unique; years later, when the Agricultural Department published a map of the distribution of banana trees in Uganda, it coincided with the distribution of cases of human bladder cancer which in Uganda was not predominantly associated with schistosomiaso.

It was due in part to studies in Uganda that the realization grew that cancer in humans is in very large part an environmentally determined disease or group of diseases. It was indeed in Africa that the late Dr. George Oettle remarked that, based on African experience, 80% or more of human cancer is environmentally determined and theoretically can be prevented.[4] Surveys in Uganda[5] showed that age and cancer dissociated, as has been found now elsewhere, thus suggesting that the long-known association with age represented the cumulative exposure to carcinogens in a particular environment. Studies of the types and sites of cancer in Uganda showed a constancy in the cancer pattern lasting more than half the century between Uganda's first contacts with the West and its modernization before its recent descent into barbarism.

That environment is so important is true of both the superficial as well as the deep internal cancers, as Kennaway noted in 1944.[6] It equally well follows that cancer epidemiology must change from superficial to deep analysis of the circumstances in which human cancers develop. Thus far, however, cancer epidemiology has singularly failed to adjust to the ideas and needs of these deeper analyses required, analyses which will be badly needed if those so well-equipped with computer skills are to have the necessary material to work on. It is beyond my knowledge and ability to discuss these tools, but I wish to look at some of the increasing complexities of exploring the environmental circumstances in which human cancers develop, and some of the basic problems in what I have elsewhere termed "the newer epidemiology of cancer."

To give us encouragement, let us envisage the potential benefits of such work. Doll[7] considers that the best comparisons between cancers in different countries or communities are made by comparing the rates per 100,000 per annum for persons between the ages of 35-60 years, thus disregarding the distorting influences that an excess of the young or old can have in certain countries. If we consider his published figures, we can see the possibility of populations with a negligible number of cancer cases. Doll's populations with extremely low figures of incidence of all the different types of cancer represent peoples who protect themselves, or are protected, from cancers occurring with high frequencies in other populations. To find how this is achieved is going to be a difficult task, but the sooner we set about it, the better.

No better example of this, and of this newer epidemiology of cancer, can be cited than the discovery, made in Boston, of the relationship of di-ethyl stilbestrol to the vaginal adenocarcinomas occurring in girls and young women.[8] The specificity of the cancerous lesions produced should be noted and should stress the need to study histologically specified cancers in specific organs, thus notably increasing the need for specific histological diagnosis. But in the study of the young females with vaginal adenosis and adenocarcinoma, study of the affected individuals themselves — however comprehensive — would not have elicited the cause; the causative agent had been administered to the mothers of the victims during the pregnancy that produced the victim. Thus, to discover the cause, the inquiry had to be extended beyond the victim to the family involved, and into the environment of the mother and the victim perhaps two decades or more antecedent to the development of the cancer. Nor is this all. The usual cancer registration practices — recording the name and address of the cancer patient — obscure etiologic relationships. It is evident that the factors that produce cancers are embedded far back in the life of the victim; to uncover them, we must begin our inquiries as soon as the diagnosis is made and the patient can be questioned.

[1]Davies, J.N.P. and Davies, A.G.M. "Jaw Tumors in Uganda Africans." Paper to Eighth International Cancer Conference, London, July, 1958. Published *Acta Unio Contra Cancer*, 1960, 16, 1320.

[2]Davies, J.N.P. and Wilson, B.A. "Cancer in Kampala." *East African Medical Journal*, 1954, 31, 394.

[3]Davies, J.N.P. and Burkitt, D.P. "The Lymphoma Syndrome in Uganda and Tropical Africa." *Med. Press.*, 1961, 245, 367.

[4]Oettle, A.G. *J. Natl. Cancer Inst.*, 1964, 33, 383.

[5]Templeton, A.G. *Tumours in a Tropical Country*. Springer Verlag. New York, 1973.

[6]Kennaway, E.L. *Cancer Research*, 1944, 4, 571.

[7]Doll, W.R. *Brit. J. Cancer*, 1969, 23, 1.

[8]Herbst, A.L.; Kurman, R.J.; Scully, R.E.; and Poskanzer, D.C. *New Eng. J. Med.*, 1972, 287, 1259.

Members of the family must also be interviewed. They are usually only too willing to help and cooperate.

When this is done, situations are uncovered that are very revealing. For some years my colleagues and I,[9] though denied any research funds, have been investigating the behavior of Hodgkin's Disease in Albany. In the last 25 years, we have had periods in which the incidence of the disease was at much the same level as elsewhere in New York State. We also had a period of unusually high incidence, followed by a period of low incidence. We not unreasonably consider this as much an epidemic as those reported with other diseases. Rather fortuitously, we discovered that a very considerable number of cases, perhaps as much as one-third in this period, were related in a peculiar way. The victims either knew cases of Hodgkin's Disease directly or were in contact with an affected person by one intermediate who remained healthy. It was possible to identify early on a group of individuals in a single high school class who either developed the disease themselves or had very close friends who did. When these contacts — who, incidentally, remained healthy — were followed, it was found that some persons who lived in the same house with them developed the disease,[10] often older men and women who, when questioned, denied personal knowledge of Hodgkin's Disease in their family or acquaintanceship. Armed with this knowledge, we tackled other school situations and uncovered the same phenomenon.

The Schimpffs in Baltimore discovered how often simple inquiry of such individuals uncovers the same situations.[11] Let us consider our experience with one elderly man, living alone in an apartment, who at age 67 develops the disease. He affirms there is no one in his family with the disease. He has limited acquaintances and they are in good health. If we, as is usual, end inquiries at this point we are baffled. But if we extend our inquiries to the man's healthy acquaintances, a totally different picture emerges.[12] This man is at the center of a whole web of Hodgkin's Disease patients, a phenomenon we find so commonly that it must be of great etiologic significance.

My purpose in discussing Hodgkin's Disease is to show you some of the non-statistical complexities we must face in cancer epidemiology. For example, where are cancer victims located? Registration is usually by home address, sometimes ludicrously by hospital. But mapping by home address may or may not be useful; the particular individual may spend a relatively small proportion of his/her time at home compared with the time spent at work or at places of play, interest, and enjoyment away from home. Moreover, home to some people is where they sleep but do little else. Thus, though we may have to pay attention to the home, to its soil, water, construction or surrounding vegetation, we may find that the carcinogens we seek may never be brought into the home. Then again, they may be, as Dr. Vianna has just demonstrated in his studies of mesothelioma in women: the wives of asbestos handlers who washed their husbands' clothing contracted asbestos and developed mesotheliomas.[13] Here the specific carcinogen was brought from a place of work.

The home can be contaminated by other means. We in New York have been concerned with contaminants airborne from polluting plants to the home. This phenomenon has been observed with asbestos; we found plastic material in the lungs of a woman living downwind from a plastics manufacturing plant; more recently we found several cases of angiosarcomas of the liver in persons who lived downwind from industrial plants which had been releasing large quantities of vinyl chloride into the atmosphere.[14] Thus, we have to consider wind and rainfall patterns in relation to certain types of cancer and as part of the environment of the home, as well as of places of work and leisure-time activity.

This, we shall have to consider, to record and to monitor what goes on in the homes of cancer victims, to inquire into diet, cooking, customs, habits and ways of life, as well as into the details of marriage, childbearing, lactation, and contraceptive use (which is done in regard to breast cancer but often omitted with other cancers). And, of course, all the time we must note the exceptional individual who manifests some singularity which marks him off from his fellows.

But if details of the home, diet, etc. are often omitted, the greatest failures in recording are in occupational histories, which seem to be a blind spot in American medical thinking and recording. It is rare in my experience to see a detailed occupational history in a patient's chart, even if the patient is suffering from a cancer notoriously often associated with occupation, or indeed, even if the patient has a well-known occupational disease. It is not uncommon to find a patient with cancer set down as "retired"; inquiry sometimes brings out that hs is a "retired parking lot attendant." Inquiry usually stops there, but further probing might reveal that, for example, the victim was a parking lot attendant when he could obtain no other employment because he was so breathless from asbestosis, a condition which can develop in many curious ways (living downwind of an asbestos plant, by a mine dump, along an asbestos-dusted highway, or through fixing brake linings or refrigerators or insulating material). I remember the occasion on which three professional men with carcinoma of the bladder in a London hospital and with no obvious occupational risks were all found, as young men, to have worked in German chemical factories handling aniline dyes, before persecution forced them to leave their country. It is too often forgotten that we may have to go far into the past to uncover what may be the decisive factors in a particular carcinogenic situation.

We must also remember, especially in interpretation, that some cancers are falling in incidence, e.g., gastric cancer of the intestinal type. Mapping by type of such falling incidence may be as important and as revealing as mapping a rising incidence. Mapping the altering pattern of cancer in immigrants, e.g., Japanese or Scandinavians, or of Chinese with nasopharyngeal carcinoma, can hardly fail to be of interest. A further field of interest I would like to suggest might be the mapping of what I call "doubles." For example, in children there seems to be some curious link in time and place, at least over large areas, between leukemia and brain tumors.[15] To map these first separately and then together might be of extreme interest.

However, to embark on too many projects would only be to

[9]Vianna, N.J.; Greenwald, P.; Brady, J.; et al. "Hodgkin's Disease: Cases with Features of a Community Outbreak." *Ann Int Med.*, 1972, 77, 169.

[10]Davies, J.N.P. in Grundmann, E. and Stulinius, H., eds. *Epidemics of Hodgkin's Disease in Current Problems in the Epidemiology of the Cancer and Lymphomas.* Springer Verlag. New York, 1972.

[11]Schimpff, S.C.; Brager, D.M.; Schimpff; et al. *Ann Int Med.,* 1976, 84, 547.

[12]Vianna, N.J.; Greenwald, P.; Davies, J.N.P. in Lacher, M. *Hodgkin's Disease.* J Wiley. New York, 1976.

[13]Vianna, N.J. and Polan, A. *Lancet,* 1978, 1, 1061.

[14]Brady, J.; Liberatore, F.; Harper, P.; et al. *J. Natl Cancer Inst.,* 1977, 59, 1383.

[15]Davies, J.N.P., in Marsden, H.B. and Steward, J.K., eds. *Tumours in Children.* Springer Verlag. New York, 1968.

find that you would be balked by lack of detailed social, chemical, histologic, occupational and other data. To begin to map cancer fully, despite excellent starts, we must have much better, and much more, basic data about patients with cancer. There has got to be a reorientation in our thinking about cancer. Once the diagnosis of cancer is made in an individual, immense interest and concern is invested in treatment, care and prognosis.[16] The victim is followed "up" until the case is closed but there is hardly any follow "down," that is, inquiry into why he or she developed his or her particular cancer. Yet, if we are to turn our now-firm evidence of the importance of environmental factors in producing cancer into really preventing cancer, not just diagnosing it early, there must be detailed "follow down." Information has to be gathered that is vital to cancer epidemiology, even though it is totally immaterial to the treatment and care of the patient.

On the whole, the medical profession is relatively little interested in prevention, though highly motivated and interested in care. To provide you with the needed basic material for computerization and mapping, perhaps we need a new type of professional, the "follow downer" such as Dr. R. Miller has called for in Pediatric Oncology. Until this or some equivalent is developed, I fear that the medical profession is not going to provide you with the vast amount of detailed information you will request. Perhaps the consideration and resolution of the difficulty should occupy much of your consideration. Of one thing I think I can assure you: once your interest and concern is made clear, you will have a remarkable degree of cooperation from patients and public.

[16]Davies, J.N.P. *Laval Medical*, 1968, 39, 45.

American Graph Fleeting, A Computer-Holograph Map Animation of United States Population Growth 1790-1970

by Geoffrey H. Dutton

The Message

In 1790 the first census of the new United States counted about four million citizens. Nearly all were non-native Americans who lived within 200 miles of the Atlantic Ocean, occupied only about 240,000 square miles. The other ninety percent of the 48 states-to-be was known mainly by rumor.

Inevitably the rumor spread, and pursued by settlers, the Republic surged westward. In a century's time sixty million people, pushed by population pressure and pulled by the land's promise, had expanded the frontier. Then, with the growth of industrialization and little territory left for expansion, population growth focussed inward, signalling the Age of Cities.

The drama of America's expansion and urbanization is at best dimly portrayed in the statistical snapshots of the census; but graphically assembled in space as maps and viewed as a process in time, even simple statistics of human habitation can come to life. To this end population data for U.S. counties have been assembled and interpolated in space and time to produce 181 maps of American population densities, one for each year from 1790 to 1970.

The only features on these maps are those of population itself, depicted as a surface, a data terrain. In these demographic landscapes the height of the surface anywhere represents density of population in that place, and the volume enclosed between the surface and its base plane is proportional to the size of the total population.

During the eighteen decades depicted, inhabited territory grew tenfold in extent, the population increased fiftyfold and the number of people dwelling in cities multiplied by more than seven hundred. This growth is shown in one-year steps, with the viewpoint steadily shifting around the compass, starting and ending over the Caribbean Sea.

The features of these surfaces are easily grasped. Where the terrain rises, population densities increase; where it ends, oceans and wilderness begin. Its valleys show where inhospitable locales have limited human occupation of the land; its summits locate urban centers, at first barely distinguishable, but dominating the peoplescape as the twentieth century unfolds.

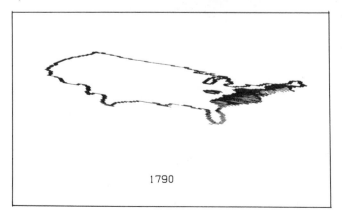

1790

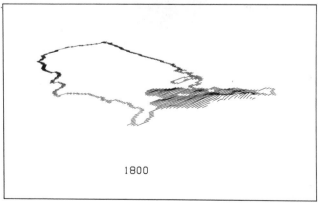

1800

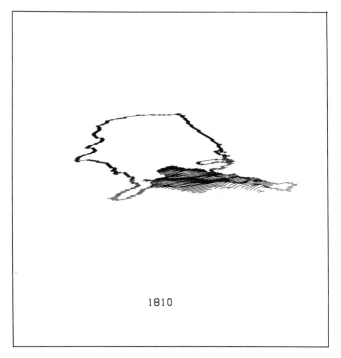

1810

1820

1830

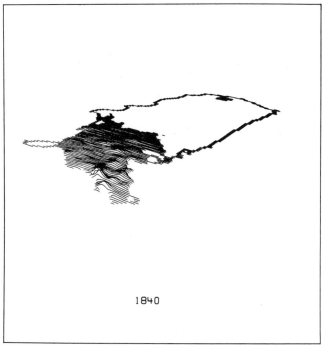

1840

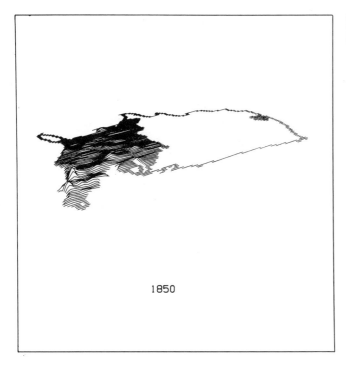

1850

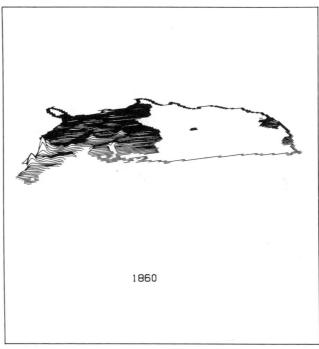

1860

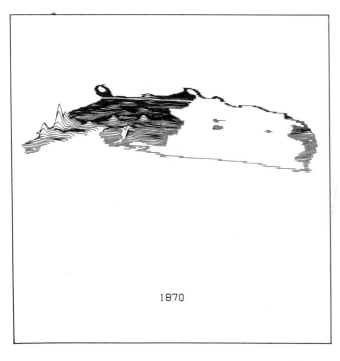

1870

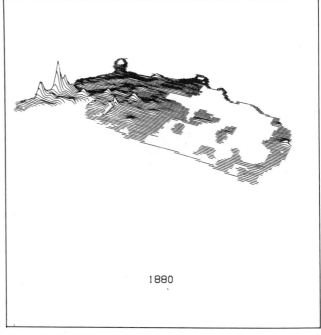

1880

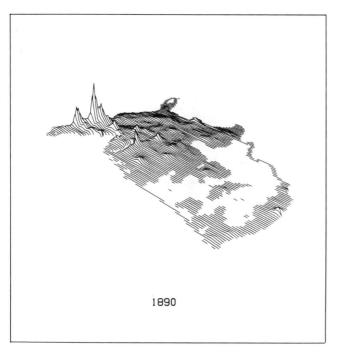

1890

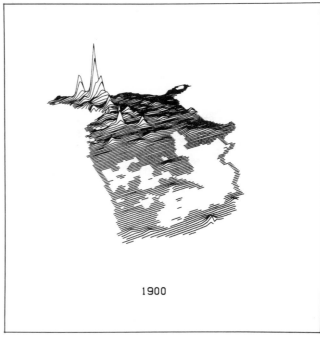

1900

1910

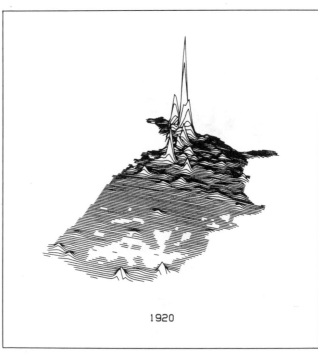

1920

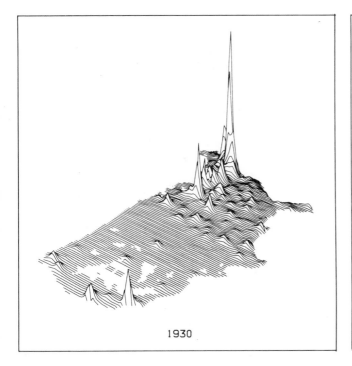

1930

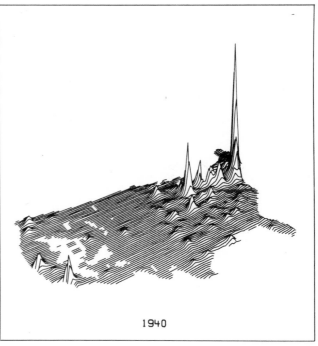

1940

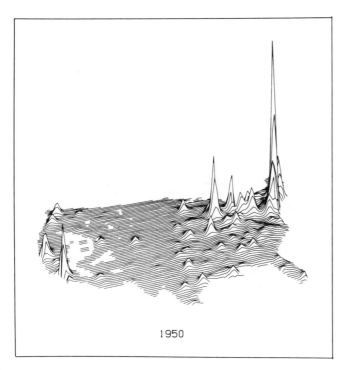

1950

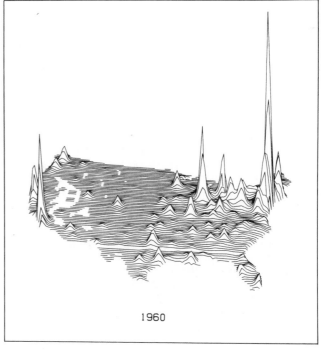

1960

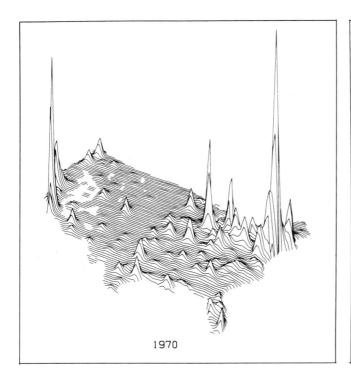

1970

Produced by   Laboratory for Computer
Graphics & Spatial Analysis,
Graduate School of Design,
Harvard University

Programming James Dougenik
Geoffrey Dutton
James Little

Film   Jody Culkin

Titles   Bruce Kennett

Holography   The Holographic Film Co.

Direction   Geoffrey Dutton

OVERVIEW

The **American Graph Fleeting** is an experimental, four-dimensional demographic map that attempts to summarize nearly two centuries of American population growth and change. It is four-dimensional in that it shows changes in inhabited land (two dimensions), population, and time. The demographic drama of the exploration, settlement, growth, and urbanization of the United States is played out on a computer-generated series of maps displayed on a cylindrical hologram. The hologram presents an animated film of 181 maps of population densities which requires neither a projector nor a laser in order to view it. Population densities are depicted as volumetric surfaces, shown in three-dimensional perspective. In these topographic-type terrain renderings, surface height indicates magnitude of population density rather than elevation above sea level.

The data presented in the display are from U.S. Census of Population by county, amounting to some tens of tousands of statistics of population densities across the 19 censuses. By generalizing these observations into smooth surfaces, the amount of detail presented could be controlled to highlight major features and damp out small local variations. In addition, whatever errors may exist in the data are conveniently masked by this approach, but the larger events and trends show

through, and to reveal these is the major purpose of the experiment.

In the beginning of the sequence (which starts with the first Federal Census in 1790) the undulations of the surfaces have appearances similar to landforms; but soon cities grow in importance, and the surface in their vicinities rises into sharp spikes, quite unlike natural topography. After 1830, New York becomes the nation's largest city, and remains the highest peak for the duration of the sequence. Other large cities, but obviously not all urban places, can be identified. Some of the earliest to appear are Boston, Philadelphia, Cincinnati and Charleston, quickly followed by St. Louis, and New Orleans. Later, places like San Francisco, Seattle, Denver and Chicago emerge from obscurity into regional dominance. Relative to the population as a whole, however, urban growth is modest until the twentieth century (not until 1920 did the Census count more persons living in cities than those who did not). The maps from about 1900 onwards reveal the intense urbanization which has characterized the twentieth century, and contrast it to the relative stagnation of rural development. Of particular note is the dramatic rise of Southern California and cities elsewhere in the "sun belt," and the emergence of megalopolitan regions in the Northeast following the Second World War.

Unlike the movie film from which the hologram was constructed, no projector or screen is needed for viewing, and any image from the film can be accessed at will. Time's tempo can be quickened, retarded, arrested, or reversed simply by changing one's viewing station relative to the display. Fast-moving events can be slowed, slow changes accelerated, and any particular region examined in detail. The ability to do so is important, as several types of action simultaneously occur. As time unfolds, the population grows and migrates away from the Atlantic coast. This both causes the surface to rise and pushes its extent westward. As these population changes occur, the viewpoint from which the surface is seen orbits about the U.S. in a great spiral, from an initial viewpoint thirty degrees in elevation above the Caribbean Sea, rotating counter-clockwise slightly with each map. By 1970 we have travelled full circle around the U.S., gradually gaining in altitude, ending up sixty degrees in elevation over Cuba. By orbiting the viewpoint, stereopsis is enhanced, and no location is permanently hidden from view by obscuring terrain. Walking around the cylinder creates the impression of circling a solid surface which continually transforms its terrain to reflect changes in time. A date is printed below each frame to identify its place in the sequence, and a narrative appears above the first half of the image sequence summarizing the action portrayed.

TECHNICAL SYNOPSIS

Three stages of work were involved in producing *American Graph Fleeting*; data base construction, map production, and holography. Each step contained a number of critical details, serially dependent on one another.

Data Base Construction

The most time-consuming portion of the project involved assembling the historical census data and transforming it to a format from which maps could be made. The starting point was a file of population counts for all 19 U.S. censuses at the county level, about 60,000 items of information. This file was acquired on tape from the Inter-University Consortium for Political Research, based at the University of Michigan. Once in hand, this file was augmented with geographic coordinates, consisting of a single point, or centroid, giving the location of the center of area of each U.S. county. This yielded a working master file, having one data record per county. On each such record are identifiers (county name and code number), centroid latitude and longitude, land area, and 19 fields of population counts, one for each census date.

The master file is not really a cartographic data base, just an archive of data. But it serves as input to a program which can construct a data base from which maps can be made. The cartographic data base takes the form of a grid of numbers covering the U.S.; each grid cell contains a population count for the area and date covered. The master file, however, contains population counts for counties, which are assumed to be located at points (their centroids). All grid cells which do not contain centroids thus have no population data, and values must be estimated for them. This process, called interpolation, proceeds as follows.

For each census date the county statistics are gridded as a matrix containing 82 rows by 127 columns. The latitude and longitude of each center is projected to planer coordinates (using an Albers equal area projection), which are then scaled to the grid dimensions. This allocates the population of each county to some grid cell. The locations of counties with respect to one another are thus roughly retained, although a few grid cells may be allocated two or three counties where counties are small. Next, a distance decay function is applied, spreading the values as given to all cells within a certain radius of each centroid cell. This radius is specifiable by the program's user, but is typically 50 to 100 miles. The distance decay function produces a tent-like, symmetrical distribution of interpolated values around each centroid, peaking there and declining outwards to a small value at the edge of the circular area affected. Thus as each county is processed, its population is spread outwards, in a probability distribution with specified extent and rate of distance decay. Interpolating all counties in this manner yields a quasi-continuous surface which is then smoothed once (using binomial smoothing operators) to remove minor peaks and any abrupt changes in value. Another grid then is referred to, providing information about the shape of the U.S. This grid is overlaid on the data grid in a way that causes all data cells covering oceans or foreign soil to be flagged as invalid. Then the sum of the remaining cells in the grid is adjusted to equal the total U.S. population at the given census date, and the interpolation process is complete.

What results from this process is a model of population distribution which, while not accurate in detail, provides a reasonable overview of the county data. Each grid cell contains a population count, which in rural areas may be as low as 500 persons, but in places like New York City may exceed a million persons. Any cell with fewer than 500 individuals was arbitrarily declared to be uninhabited (i.e., set to zero); This corresponds to a population density of one person per square mile (since each cell is a square covering 500 square miles of land or water). Note that densities of native American people in unorganized areas are not displayed, even if greater than one per square mile. This is unfortunate, as it would have been quite revealing to compare changes in settlement patterns for native and non-native Americans.

The grid cell size was chosen to yield a reasonable amount of detail without accumulating exorbitant processing costs. A finer mesh could have been used, but by making the cells half as wide, computation time would have been increased by a factor of four to six.

Map Production

Having created a data grid for each of the 19 censuses, each grid was then displayed as a perspective view of a surface, along with intermediate grids interpolated from the 19 basic ones. In such surfaces, the higher the value in a cell, the higher the surface rises above the cell's location. The program which accomplishes this, ASPEX (for Automated Surface Perspectives) allows its users to specify the angle of view, maximum height of the surface, size of the map, and other graphic parameters. ASPEX can be run interactively and can display its maps on a CRT screen, but because of the movie-making requirements of the project it was run in batch style, producing display files of graphics rather than drawings. One map per year (from 1790 to 1970) was produced, shifting the viewpoint for each view two degrees counter-clockwise, so that with the final map, the viewing azimuth would coincide with that of the first map, having travelled full circle around the U.S. In addition, the viewing altitude was gradually increased (from an elevation of 30 degrees initially to an ultimate elevation of 60 degrees).

The rationale for this was that the flatter surfaces of earlier decades are better perceived from a low viewpoint, while toward the end of the sequence, when urban peaks rise to great heights, a higher viewing angle gives more visual information. The change in altitude within any given decade, however, is barely noticeable.

Command Files

A small program was written to create command files for ASPEX, which specified the viewpoints for all 181 maps in the series. This enabled a smooth, regular transition from each map to the next, and eliminated the tedious and error-prone task of typing in commands to specify the views. ASPEX itself was augmented to derive data grids for intermediate dates, by linearly interpolating between successive census grids. Thus not only was the viewpoint for each map unique, each view also portrayed a slightly different surface from the ones that came before and after it.

In drawing the maps, ASPEX examined each cell along every row in a given data grid, calculating the height of the the surface (based on the population count in the cell, its position in the grid and the perspective characteristics of the particular view) over that cell. The surfaces were constructed by drawing parallel profiles (i.e., relief lines oriented at 90 degrees to the surface contours) along rows (or columns) in the grid. Profiles were plotted, beginning at the edge of the grid closest to the point of view, successively backward toward the rear of the grid. The pen (or beam) was directed to trace profiles across the map, comparing the height reached in each cell to a "horizon" representing the highest point reached at the same x-position on previous profiles. If the new profile was higher than the current horizon, the horizon was updated to reflect it; if it was found to go below the horizon, the pen was lifted until the profile emerged above the horizon. This is how the portions of the surface hidden from view were determined by the program and omitted from the rendering. In addition, the pen was lifted whenever the cell value was zero, except for cells lying on the boundary of the U.S.; border cells were always plotted to provide a frame of reference for the maps.

In several late-night sessions of running ASPEX, display files containing all the graphic coordinates to draw 181 maps were created. These were previewed for verification using a simple program to read them and plot the coordinates on a CRT. Once checked, the display files were written to a magnetic tape and the tape taken to a service bureau for final plotting. This strategy was employed because of the existence in the Boston area of a high-quality film plotting device, the Information International FR-80 plotter. This device exposes 35 mm film or hard copy paper through optics to a high-resolution CRT on which lines can be drawn. Since the lines to be plotted had already been calculated by ASPEX and their endpoints recorded on the display files, using the FR-80 involved only one small program similar to the one used to verify the display files, but set up to issue the appropriate instructions to drive the FR-80.

Due to the fact that the FR-80 camera cannot handle 16 mm film or 35 mm film having sprocket holes (it is a microfilm/fiche production device), hard copy (paper) output was chosen for convenience as the medium. These plots were then filmed using 16 mm high-contrast film in temporal sequence, and titles incorporated into the sequence in the process of filming. This was done using a professional animation stand having a pin-registered camera. Five frames were recorded on film for each map, plus about 150 frames for the main title and credits, resulting in a 1,080-frame film animation. This amount of film takes only 45 seconds to show when projected (at 24 frames per second), but in the static conditions of viewing the hologram the rapidity of the sequence poses no problems for the viewer other than a slight blurring.

Holography

The medium for this display, the "integral hologram," is a relatively recent extension of the rapidly maturing art and science of holography. The process permits multiple images to be recorded and directly viewed using one large strip of holographic film. The images are typically frames from movies, and the current state of the art permits about a thousand frames to be recorded on a hologram which, when bent into a cylinder, measures 16 inches in diameter and 9-1/2 inches in height. Longer sequences can be captured by using a longer strip hologram, yielding a wider cylinder. The content of the movie may be an action sequence, such as a horse race, or may simply record a rotation around an object or a person (describing a product or yielding a three-dimensional portrait). The current technology does not permit color to be captured; the virtual images appear in color however, due to refraction of the viewing light source by the hologram. These colors are spectral, and have no relationship to the color characteristics of the original objects filmed.

The process of creating integral holograms is fairly straightforward, and is diagrammed on the following page. Using a specially built filming apparatus (which must be mechanically isolated from all sources of vibration, due to the close tolerances involved), movie frames are sequentially projected using a laser for a light source onto successive adjacent vertical strips of holographic film. The width of the strips is controlled by a fine slit mask, a fraction of a millimeter in width. The laser beam is split, as it would be when illuminating an ordinary object for a hologram, into an "object beam" and a "reference beam." The object beam passes through a movie frame, and then through optics to the holographic film plane. There, within the slit mask, the two beams converge, creating patterns of interference which record that frame as a hologram. The entire movie is thus captured serially across the rectangular hologram as a thousand very narrow holograms.

Upon developing it, the hologram is bent into a cylinder and illuminated from below by an incandescent light source. The light source is positioned at the same angle relative to the film that the reference beam occupied when exposing the hologram. Because of the cylinder's curvature and other critical angles of the process, only a few frames are visible at once to a viewer. A sufficient number are visible, however, to generate a stereoscopic sensation of depth, so that the virtual image seen appears in three dimensions, floating inside the cylinder of film. This image is reproduced as a negative, with light lines comprising the surface, which floats in a dark void. If the cylinder is rotated or the viewer walks around it, frame after frame comes into view, animating the images.

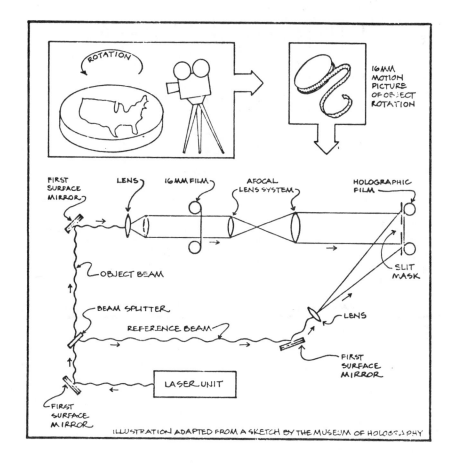

ILLUSTRATION ADAPTED FROM A SKETCH BY THE MUSEUM OF HOLOGRAPHY

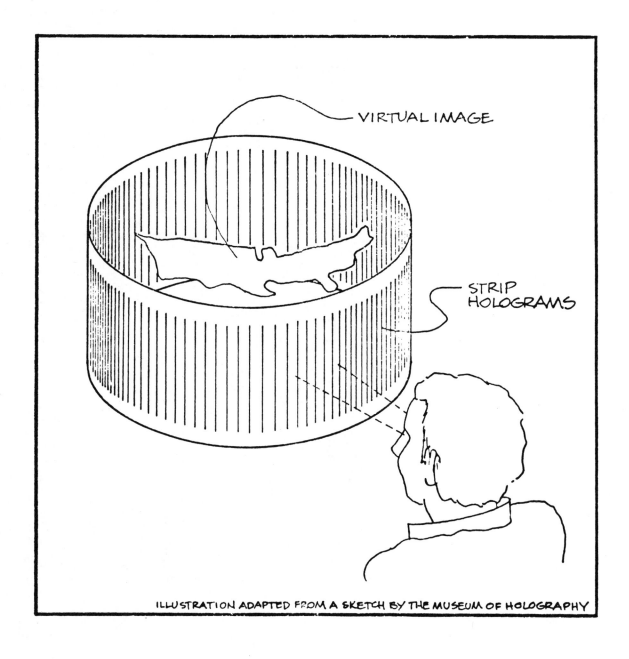

VIRTUAL IMAGE

STRIP HOLOGRAMS

ILLUSTRATION ADAPTED FROM A SKETCH BY THE MUSEUM OF HOLOGRAPHY

SYMAP as an Aid to Teaching Thematic Cartography

by Alun Hughes

BACKGROUND

This paper describes and evaluates the use made of computer graphics, specifically SYMAP, in teaching cartography to geography majors at Brock University. The course in question is GEOG 302, Cartology, which is taught over a 14-week period in third year and which is preceded by an introductory cartography course in year two. The primary theme of GEOG 302 is "the map as a medium of communication," with special reference to thematic maps. In developing this theme, a conscious effort is made to identify the key decisions which enter into the map-making process, to examine their consequences, and to discuss their implications for the map-maker and the map-user. It is felt that an appreciation of this decision-making aspect is crucial to a proper understanding of the map as a medium of communication.

The compilation of an isopleth map (the main type of map produced in GEOG 302) affords a good illustration of this. Commencing with a set of area-based data values (e.g., population density figures for a group of counties or some other type of data zone), the cartographer first assigns each value to a unique control point within the corresponding data zone. Then, on the assumption that the control point values are sample values taken from a continuous statistical surface, he maps the surface by interpolating isopleths between the control points. If we analyze first the traditional hand method of isopleth mapping (Mackay, 1951; Schmid and MacCannell, 1955; Porter, 1957-58; Hsu and Robinson, 1970), we see that the cartographer is called upon to make five, and possibly six, distinct decisions: where to place the control points within the data zones, what type of gradient to assume between control points, which pairs of control points to connect to form a triangular network of interpolation axes, how to divide up the data set into class intervals, how to extrapolate the isopleths to the map margin, and (optionally) what sequence of tints or colours to employ in layer shading. In the special case where the control points are arranged on a square grid, the question of how to resolve the alternative choice problem (Mackay, 1953) replaces the decision on interpolation axes.

Each of these decisions has an effect on the appearance of the mapped distribution and ultimately on the map-user's interpretation of that distribution. Though sometimes the effect can be relatively minor (e.g., the displacement of isopleths due to the shifting of control points from centers of gravity to centers of population), it can also be quite spectacular, as for instance when class intervals are changed (Jenks, 1963), or when the normal sequence of layer shading is inverted (in other words, using the analogy of hypsometric shading, when the progression is changed from "the higher the darker" to "the higher the lighter").

Each of these decisions is also essentially subjective. Despite the ever-burgeoning volume of modern research, there is yet very little in the way of "cartographic law" to guide the cartographer toward the "correct" solution. There are of course certain cartographic conventions that have emerged over time and which ostensibly remove the burden of at least some decision making from the cartographer's shoulders. Thus the assumed gradient is almost always a linear one and interpolation axes are generally drawn between control points only if the corresponding data zones touch. But these conventions are not laws and the cartographer retains the option of choosing from any number of alternative solutions should he so wish. Indeed, it is important that this option remain, for there is no certainty that the conventional solutions are also the most desirable from the theoretical standpoint. At this stage in the evolution of the discipline, with cartographic theory still very much in the developmental phase, it is simply not possible to prescribe "correct" solutions to many map-making problems. Thus, though the conventional solution to the problem of selecting interpolation axes is much simpler than the "minimum length" solution advocated by Yoeli (1977),[1] there is no guarantee that it is also the theoretically sounder of the two. The existence of conventions does not, therefore, absolve the cartographer from the need to think for himself.

Where some decisions are concerned there are no conventional solutions anyway. The selection of class intervals is a case in point. The literature abounds with guidelines on how to select the appropriate number and sequence of intervals, but these are often vague or even contradictory. As Scripter (1970, p. 387) puts it with nice understatement: "The state of the science of class interval selection is rather rudimentary," and in the end the cartographer has no alternative but to decide for himself.

It follows that an isopleth map is never *the* definitive representation of a given distribution. There can be no such thing. It is only one of numerous possible representations, any one of which might have been produced had the cartographer made different decisions (including, of course, the choice of another mapping method — choropleth mapping, for example). It is clearly important that the student be aware of this fact, whether he is producing an isopleth map himself or reading an isopleth map produced by someone else.

It might be argued that the foregoing example — isopleth mapping by hand — is no longer very realistic, since isopleth maps are nowadays almost invariably produced by computer. Not only does the computer often employ a completely different interpolation model, but the computer, it is said, largely eliminates the human decision-making factor. However, a moment's reflection shows that the same considerations apply whether the map is produced by hand or by machine. The fact that a variety of interpolation models exists in no way confounds

[1]Though Yoeli's solution was developed in the context of isometric mapping, there is no reason why it could not be applied to isopleth mapping.

the previous discussion; quite the opposite, for the choice of model simply represents yet another decision that has to be made (Rhind, 1971; Morrison, 1974). And though the existence of defaults permits the cartographer to by-pass certain decisions (e.g., class intervals, symbolization), there is always the possibility of overriding them. Defaults are the machine equivalent of conventions and the same arguments apply as before. As Liebenberg (1976) forcibly reminds us, computer-drawn maps are as much a function of subjective decisions as their hand-drawn counterparts. Unfortunately, computers radiate such an aura of impersonal objectivity that it is easy for the uninitiated to be deceived into believing the opposite. In the words of McCullagh and Sampson (1972, p. 117), "Many believe that a contour map produced by a computer is somehow sacred, and must be ojective." For this reason, it is now doubly necessary that the student be made aware of the subjective element in the mapping process.

If the computer increases the need, it also provides the means by which the need can be satisfied. No amount of lecturing can substitute for the student producing maps and investigating the effects of making different decisions himself. In the past, the slowness of the hand-mapping process made this next to impossible to achieve, but the computer has changed all that. Given a study area and a suitable data set, it is now very easy for the student to run off map after map on the computer, each time varying the decisions he makes. This is essentially what is done in the practical work in GEOG 302 described later in this paper.

The practical work is also designed to enable the student to investigate an important pre-mapping decision. This is the partitioning of the study area into data zones. Normally, of course, the choice of data zones is not the cartographer's to make, since the data sets he maps are usually collected by others and are already aggregated by data zone. The type example is census data, which until relatively recently has been obtainable only for standard geostatistical areas such as census tracts, the boundaries of which are fixed by the census agency.

Nevertheless, the map-maker (or map-user for that matter) is ill-advised to overlook the significance of the data zone partition. Any isopleth map is a representation of a distribution in the real world; in the case of population density, for example, the distribution is one of individual human beings over the earth's surface. The data set of density values from which the map is made is but a generalization of this distribution, a generalization conditioned by the number, shapes, sizes and orientations of the data zones. Different data zone partitions yield different data sets, different maps, and ultimately different perceptions in the mind of the user.

A difficulty in studying the effect of partitioning in the classroom is that a real-world distribution is needed to serve as a starting point, but existing data are usually available only in aggregate form and the time factor precludes the collection of original data. The solution adopted at Brock is to create a hypothetical real-world distribution. The method of doing so owes a major conceptual debt to the geographical information system developed at Statistics Canada for the manipulation of

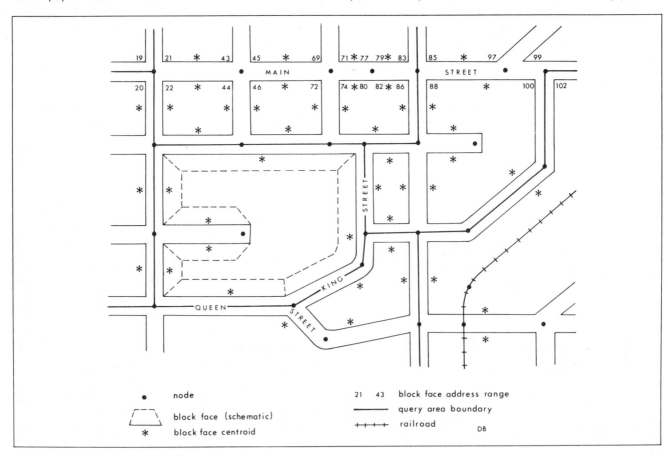

Figure 1. Components of the GRDSR Area Master File

1971 census data, namely the Geographically Referenced Data Storage and Retrieval (GRDSR) System (Statistics Canada, 1972).

The GRDSR System operates on data which are identified by street address, such as census data and municipal assessment data. Among other things, it permits the aggregation of such data to user-defined data zones called query areas, thereby freeing the user from the straitjacket of standard geostatistical areas and other systems of fixed data zones.

Supplying the spatial framework of the System is the geographical base file or Area Master File (AMF). In urban areas the basic components of the AMF are threefold (Figure 1):

1. the UTM coordinates of every node, i.e., a point at which a linear feature such as a road or railway changes direction abruptly, terminates or intersects another such feature;

2. the street name and address range of every block face, i.e., one side of a street between successive intersections;

3. the UTM coordinates of every block face centroid, i.e., a point recessed 22 m from the center of the road at the mid-point of a block face.

The AMF facilitates two important data manipulations: geocoding and point-in-polygon aggregation. By the process of geocoding, each data record containing information pertaining to a household, person or facility is assigned to a block face centroid. In effect, what happens is that the original spatial identifier for the record, its street address, is replaced by a new identifier, the UTM coordinates of the appropriate centroid.

By the point-in-polygon process, the values attached to centroids are aggregated to query areas defined by the user. To permit this, all query area boundaries must be reduced to straight line segments and the UTM coordinates of the vertices measured. By comparing vertex and centroid coordinates, the computer allocates centroid values to query areas and accumulates the desired aggregate values. The query areas are obtained (ideally at least) by combining complete block faces; for this reason the block face is commonly referred to as the "building block" of the GRDSR System.

METHOD[2]

The practical work in GEOG 302 takes the form of a series of linked assignments, each of one week's duration, involving extensive use of the computer. Background information (e.g., description of the GRDSR System, discussion of isopleth mapping theory, explanation of the SYMAP interpolation model) is provided as part of the lecture course. The work varies somewhat each time the course is taught, and the description which follows refers to the Spring term, 1975. The work was divided into three phases:

1. Creation of Hypothetical Distribution (Assignments 1 & 2)
2. Aggregation of Data (Assignments 3 & 4)
3. Mapping (Assignments 5-9).

Creation of Hypothetical Distribution

The class comprised 15 students and was divided into five

groups of three. A study area in north St. Catharines was outlined on a city street map and was divided into five sections. One section was assigned to each group (Figure 2). On separate gridded overlays, each group plotted the 250 or so block face centroids within its section and measured their coordinates to the nearest 0.01 inch. The centroid locations were estimated visually and the coordinate measurement was done manually.

Arbitrary values, intended to represent the number of people living on each block face, were then assigned to the centroids. Various methods of doing this were employed by the different groups, some systematic, some random, the aim being to create a population distribution with a general trend from high density in the northeast to low density in the southwest and also with a reasonable amount of local variability.

One computer card was punched for each centroid, containing its coordinates and its "population" value. The cards for the whole class were combined and placed on disk, at which point the hypothetical distribution was considered complete. Note that this distribution, being aggregated to the block face level, was not strictly speaking a true real world distribution. Given the method of creating data zones, however, it could be regarded as such for all practical purposes. This is explained below.

Aggregation of Data

The next step was for each group to subdivide the study area into 30 data zones, observing certain rules about data-zone shape and size in the process. The rules were as follows:

Group A: shape — variable; size — variable
Group B: shape — elongated east-west; size — constant
Group C: shape — compact; size — increasing from north to south
Group D: shape — elongated north-south; size — constant
Group E: shape — variable; size — variable

Figure 3 shows the data-zone partitions obtained by each group.

An additional constraint was that the data-zone boundaries should, as far as possible, follow street lines and avoid splitting block faces. This made for considerable difficulties in fulfilling the requirements about data-zone shape and size. But this practical drawback was offset by a significant conceptual gain. Since the resultant data zones were composed almost without exception of complete block faces, the fact that the hypothetical distribution had been aggregated to the block face level became immaterial. For all practical purposes, the distribution on disk *was* the unaggregated real world distribution.

The coordinates of the data-zone vertices were measured and the information fed to two computer programs: (1) AREA,[3] which computed the ground area of each data zone; (2) PTPL, a point-in-polygon routine which assigned values from the block face centroid file stored on disk to data zones and calculated a total population value for each.

By dividing the area into population, each group finally obtained 30 population density values, which constituted the data set to be mapped. Clearly, since the groups had employed different data-zone partitions, the five data sets differed appreciably. Thus, Group A's values varied from a minimum of 0 to a maximum of 14,651, and 22 of the 30 values were smaller

---

[2]This section is based largely on the section called "The Practical Work in Geography 302" in Hughes, 1976.

[3]Listings of this and other programs mentioned in the text are available from the author.

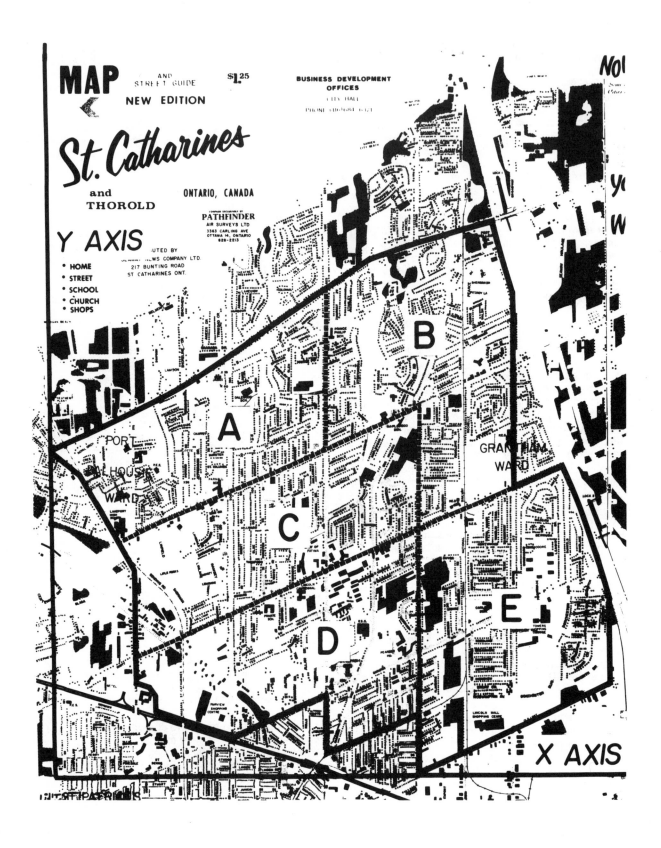

Figure 2. Base map and study area divided into sections (reproduced by permission of Pathfinder Air Surveys Ltd.)

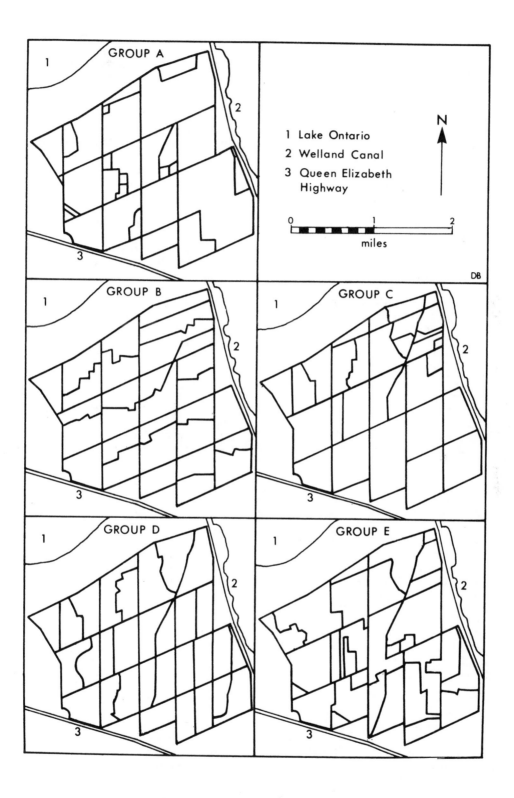

Figure 3. Data zone partitions created by each group

than 5000. For Group C, the range was 65 to 20,205 with only 11 values below 5000.

Mapping

All maps were run on the university's Burroughs 5500 computer using the SYMAP program. SYMAP is a sophisticated mapping system developed at the Laboratory for Computer Graphics and Spatial Analysis, Harvard University. It produces six types of map on a standard lineprinter: contour (i.e., isoline), conformant (i.e., choropleth), proximal, base, trend surface and residual (of which only the first three were available at Brock in 1975). In addition, it permits considerable flexibility of map output through the provision of almost 40 options or "electives" (Dougenik and Sheehan, 1975). Each group produced 15 maps in all, and these were run in a controlled sequence in order to investigate the effects of the following decision variables on the appearance of the distribution: map type, data zone partition, number of class intervals, sequence of class intervals, number of control point values used in interpolation, and symbolization.

The maps were displayed side-by-side on tackboards to facilitate comparison.

In the first mapping assignment (i.e., Assignment 5), the variable was map type. Each group produced three maps, choropleth, proximal and isopleth (Maps 1-3, Figure 4). No electives were specified (except the obligatory electives for the proximal map), which meant that scale, class intervals, symbolization and so on were established by default. For this reason the maps were, strictly speaking, comparable only within groups. Between-group comparisons were not valid because of class interval differences. These came about as a result of two factors: first, the different data sets used by each group and, second, the method used by SYMAP to determine class intervals, which automatically gives five intervals derived from the particular data set submitted for mapping.

In Assignment 6, certain variables were held constant in order to investigate the effect of data-zone partition and number of intervals. This was achieved by producing only one type of map — isopleth — and by using Electives 3, 4 and 5 to standardize class intervals. Each group again produced three maps, with 5, 7 and 9 constant intervals (Maps 4-6, Figure 5). Within-group comparison showed the greater detail obtainable with a large number of intervals, while between-group comparison revealed the effect of different data-zone partitions: the general pattern was the same for each group, but appreciable differences in detail — e.g., locations of maxima and minima — were evident.

For Assignment 7, the map type remained isopleth and the number of intervals was kept constant at 7, but the sequence of intervals was varied. Three maps were produced by each group, using Elective 6 to give arithmetic, geometric and harmonic intervals; in each case the first term of the generating series was 1 and the common difference or common ratio was 3 (Maps 7-9, Figure 6). An internal calculating program, IVAL, was run as a check on accuracy. Since the intervals were no longer standardized between groups, only within-group comparison of the different interval sequences was possible. Not unexpectedly, this variable was responsible for the most spectacular variations in pattern.

In Assignment 8, an attempt was made to manipulate the interpolation model by employing Elective 36. A brief word is necessary here about the basis of this model (Shepard, 1968).

Interpolation in SYMAP is normally a two-stage process. First the program calculates values for a grid of printing cells comprising every third cell across the map and every second cell down the map. Then it calculates values for the intermediate cells by simple linear interpolation. Only the first stage is of interest here.

In calculating a value for a particular grid cell, SYMAP centers a circle of predetermined radius on that cell and counts the number of control points falling inside the circle. If the number is between 4 and 10, it proceeds to calculate the weighted mean of the values at these points, the weights being the functions of distance and direction to the cell. This mean value is then assigned to the cell. If the number is less than 4 or greater than 10 the circle is enlarged or reduced until it does contain 4 or 10 control points, and then the calculation is performed.

It follows that interpolation is always based on a minimum of 4 and a maximum of 10 (an average of 7) control points. Elective 36 permits the user to vary these minimum and maximum values. In the GEOG 302 assignment, three maps were run by each group, the first with a minimum of 0 and a maximum of 2, the second with a range of 4 to 6, and the third with a range of 8 to 10 (Maps 10-12, Figure 7). The class intervals in this (and the final) assignment were the arithmetic intervals generated in Assignment 7. Comparing the maps within each group, the effect of increasing the number of control points used for interpolation was seen to be a progressive smoothing of the surface.

In the ninth and final assignment, Elective 7 was used to vary the printing symbols. Again three maps were produced per group (Maps 13-15, Figure 8). The first had black instead of white isopleths; the second had black isopleths with no shading in between; and the third had white isopleths, black control points and the reverse of the standard interval shading. Given that it is the dark areas of a black-and-white map that tend to attract the eye, the comparison of Maps 13 and 15 was particularly instructive.

No final report was required, since the students also had a major essay to complete. However, they were told in advance that the following compulsory question would appear on the final exam: "Describe the experiment carried out during the term to investigate the subjective element in isopleth mapping, and discuss the implications of your findings."

EVALUATION

The approach employed in GEOG 302, exploiting the power and flexibility of the computer, provides the student with a vivid demonstration of the subjectivity of the isopleth mapping process. A large number of maps is produced, and, though every map depicts exactly the same real world situation, each one is different. They are run in a highly systematic manner, which enables the student to see clearly the effects of the various decisions the cartographer has to make. Though the maps produced are mostly isopleth, the decisions involved are similar or analogous to decisions required by other mapping methods, so the lessons learned have general application. Hopefully, the student gains new insight into the nature of thematic maps and views them with a much more critical eye thereafter.

The approach is easily transferable; all that is required is a computer large enough for SYMAP (or some equivalent program) and a fair amount of display space. It is also easily modified to suit different requirements and circumstances. For example, on the two occasions that GFOG 302 has been taught

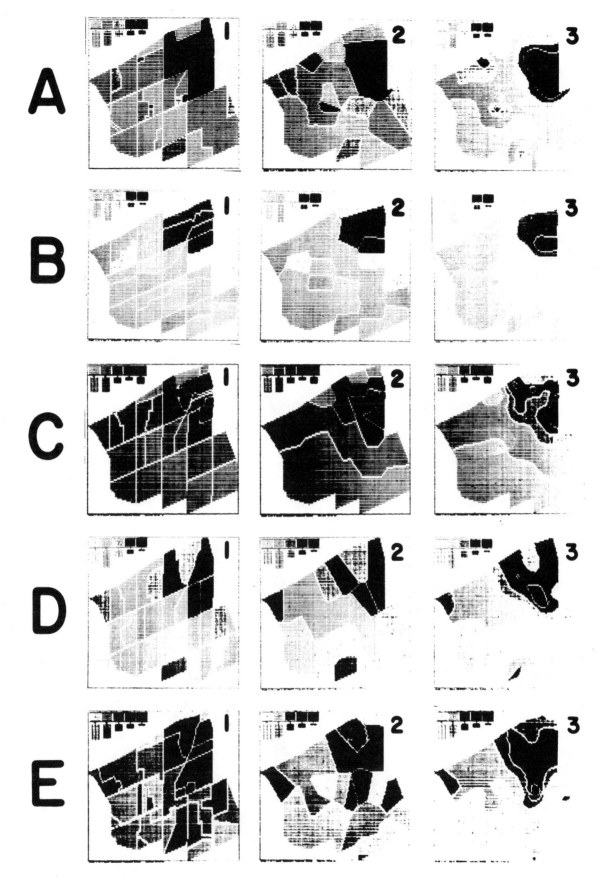

Figure 4. Maps run in Assignment 5
Decision variable: map type (choropleth, proximal, isopleth)

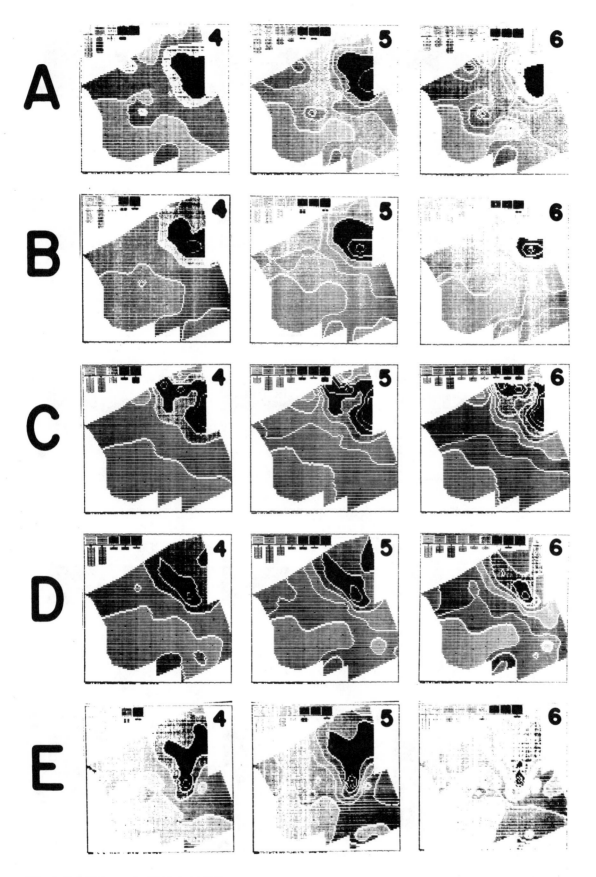

Figure 5. Maps run in Assignment 6
Decision variables: between groups — data zone partition (as Fig. 3) within groups — number of intervals (5, 7, 9)

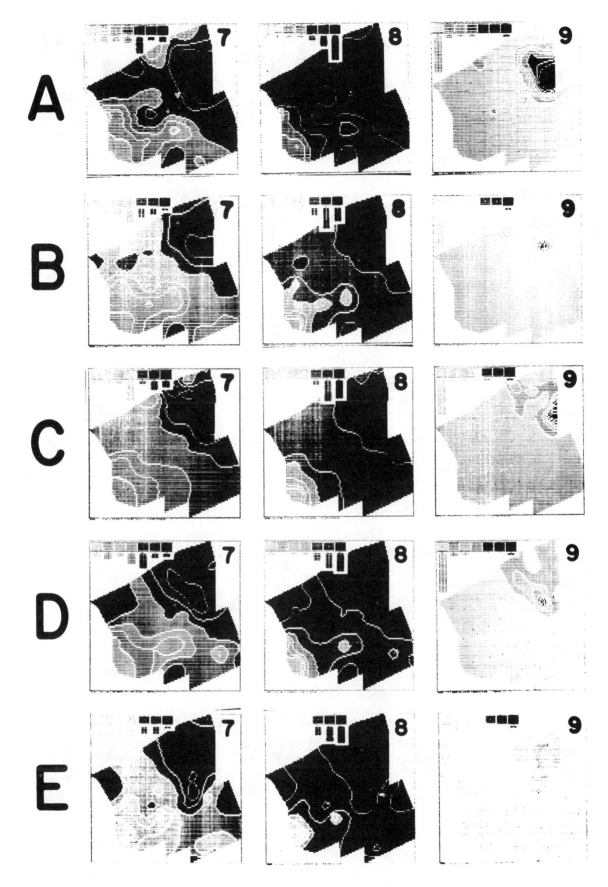

Figure 6. Maps run in Assignment 7
Decision variable: sequence of class intervals (arithmetic, geometric, harmonic)

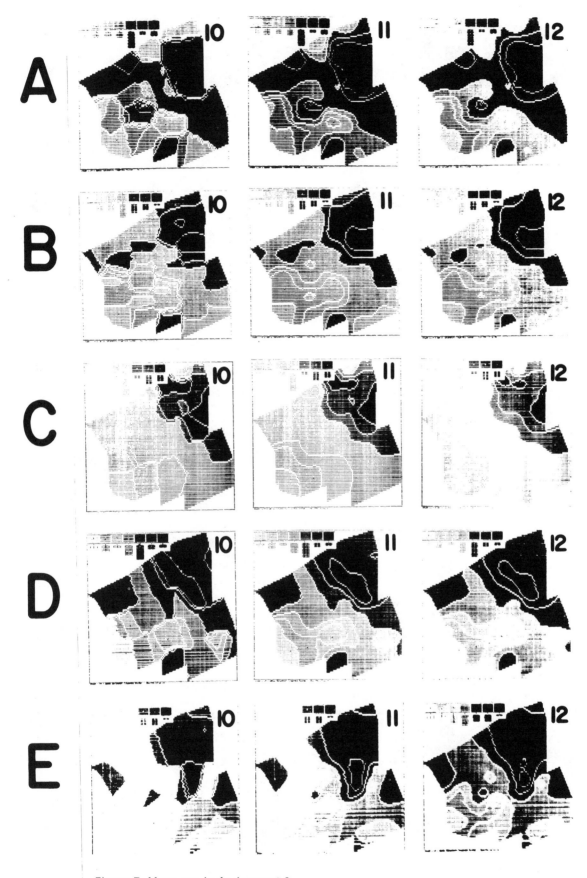

Figure 7. Maps run in Assignment 8
Decision variable: number of control point values used in interpolation (0-2, 4-6, 8-10)

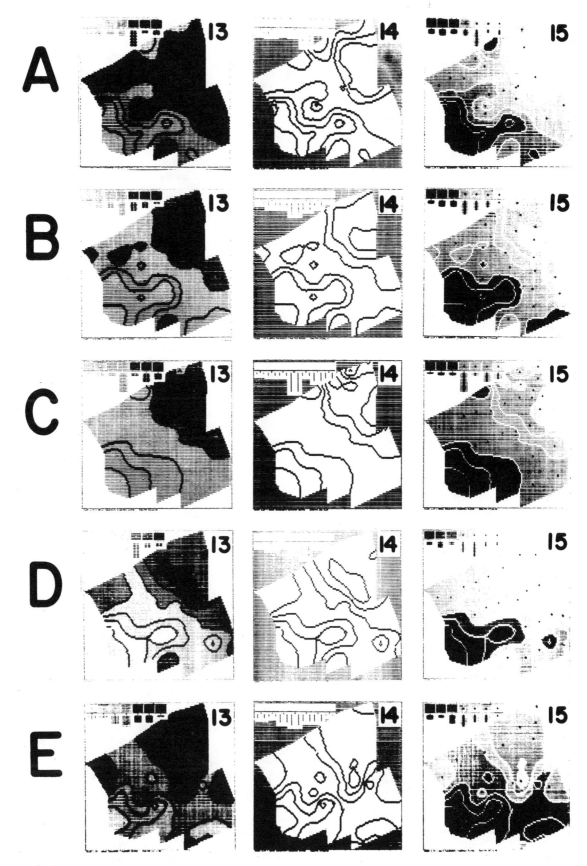

Figure 8. Maps run in Assignment 9
Decision variable: symbolization (black isopleths, black isopleths and no shading, reverse shading and black control points)

in the more restricted time frame of the Summer Evening session, the class has commenced work at the Aggregation of Data stage. To facilitate this, the block-face centroid file prepared by the previous day class was simply re-used. Similarly, in the most recent session (Spring term, 1978), an extra mapping assignment involving the trend surface and residual options (Elective 38) was introduced. There remain other possibilities for investigation — the D-BARRIERS option and Elective 37: Interpolation Grid are examples that immediately come to mind. Another possibility would be to produce isometric rather than isopleth maps, and to investigate the effect of different control point samples instead of different data-zone partitions (Morrison, 1974).

The approach is not without its problems however, and consideration of these is an essential part of any evaluation.

First, the approach requires a great deal of organization on the part of the instructor to ensure that everything goes smoothly, especially in the relatively complex initial stages prior to mapping. Since the work is cumulative from week to week, since each group contributes in some measure to the overall class effort, and since most students are quite unfamiliar with the computer, it is necessary to plan the work very carefully, to spell out instructions in great detail and to institute checks on accuracy whenever possible. Experience has shown the latter to be particularly important for two reasons. First, locating and rectifying errors causes delay and frustration, especially if they are not detected until some time after they occur. On one occasion, for example, inconsistencies which emerged only at the point-in-polygon stage were found to have originated when some groups slightly mislocated the coordinate axes on their original base maps. Second, if errors go undetected the final stage of map comparison is undermined. For example, the variations exhibited by Maps 7, 8 and 9 are supposed to be due to one factor only: different class interval sequences. If the variations are also due in part to an unknown error component, the comparison of these maps is obviously less than valid.

A second problem is that the workload is not consistent throughout the course. The initial stages involve the student in a fair amount of time-consuming work, but the workload drops off sharply once mapping starts. While the one tends to balance out the other, it does seem desirable to cut down the initial work as far as possible, especially as much of it has little intrinsic merit (e.g., plotting, coordinate measurement, keypunching). One possibility would be to use a digitizer for coordinate measurement. This was the intention for the most recent session, but since the university was in the process of installing a new Burroughs 6700 computer it seemed wiser to wait. Instead, the work load was reduced by substituting an imaginary urban area, comprising 96 identical square blocks with centroids already plotted, for the usual road map base, and by presenting each group with a pre-determined partition consisting for the most part of data zones of constant shape and size.

Unfortunately, this only served to exacerbate another problem — the fact that the approach in its present form tends not to foster student initiative. It is not a little ironic that an approach that seeks to instill in the student an appreciation of decision making in thematic mapping allows him to make relatively few decisions himself. There is, of course, a good reason for this — the need to ensure the strict comparability of maps — but hindsight suggests that this is perhaps not a sufficient reason for putting students in a straitjacket. Happily, the approach can be easily modified to accommodate greater student choice, even if this means sacrificing a degree of comparability and comprehensiveness. The computer-based instructional system described by Baumann (1975) is a possible model to follow.

There are also the usual problems which arise when students work in groups, especially over a long period of time. If the group is too large (three is borderline), there will always be those who coast along while others do the work and, worse, who soon lose track of what is going on. Two is probably the ideal size, *if* the computer budget can stand the cost of the extra maps this entails.

Cost is an obvious problem in its own right. At the moment it is not a consideration at Brock because the capacity of the B 6700 exceeds demand and there are no restrictions on departmental use. But the running of so many computer maps in such a short period of time could pose difficulties at many installations, especially when allowance is made for the number of maps that have to be re-run because of errors.

A final problem, hardly a predictable one, is the fact that some students have completed GEOG 302 with disconcertingly hazy ideas about how to run SYMAP! This is probably because the complexity of the initial assignments, combined with the strains of getting used to the computer, caused them to lose sight of the logic of data preparation for SYMAP. This drawback has now been eliminated by introducing a simple computer mapping assignment, involving the production of choropleth, isopleth and proximal maps, right at the beginning of the course.

Though problems exist, they do not alter the fact that the approach followed at Brock is both a valid and a fruitful one. As far back as 1971, Hsu and Porter (p. 799) warned of the danger that "geographers may unthinkingly accept mapping programs without understanding their underlying logic" and pointed to "abuses of computer mapping (which) have produced meaningless maps in large quantities." By using computer cartography to teach cartography, the Brock approach helps to reduce the danger and to forestall the abuses. Hopefully, it goes further and develops in students a much more responsible attitude towards maps in general and immunizes them against the affliction which Boggs (1947, p. 469) calls "cartohypnosis," which causes its sufferers to "accept subconsciously and uncritically the ideas that are suggested to them by maps."

BIBLIOGRAPHY

BAUMANN, PAUL R. (1975) "A Computer-Based Instructional System for Maps," *Journal of Geography*, 74, 3, 159-166.

DOUGENIK, JAMES A. and DAVID E. SHEEHAN (1975) *SYMAP User's Reference Manual*, Laboratory for Computer Graphics and Spatial Analysis, Harvard University, Cambridge.

HSU, MEI-LING and ARTHUR H. ROBINSON (1970) *The Fidelity of Isopleth Maps*, University of Minnesota Press, Minneapolis.

HSU, MEI-LING and PHILIP W. PORTER (1971) "Computer Mapping and Geographic Cartography," Review Article, Association of American Geographers, *Annals*, 61, 796-799.

HUGHES, ALUN (1976) "An Application of Computer Cartography in the Teaching of Cartography," *Canadian Cartographer*, 13, 139-157.

JENKS, GEORGE F. (1963) "Generalization in Statistical Mapping," Association of American Geographers, *Annals*, 53, 212-231.

LIEBENBERG, ELRI (1976) "SYMAP: Its Uses and Abuses," *Cartographic Journal*, 13, 26-36.

MACKAY, J. ROSS (1951) "Some Problems and Techniques in Isopleth Mapping," *Economic Geography*, 27, 1-9.

MACKAY, J. ROSS (1953) "The Alternative Choice in Isopleth Interpolation," *Professional Geographer*, 5, 4, 2-4.

McCULLAGH, MICHAEL J. and ROBERT J. SAMPSON (1972) "User Desires and Graphics Capability in an Academic Environment," *Cartographic Journal*, 9, 109-122.

MORRISON, JOEL L. (1974) "Observed Statistical Trends in Various Interpolation Algorithms Useful for First State Interpolation," *Canadian Cartographer*, 11, 142-159.

PORTER, PHILIP W. (1957-58) "Putting the Isopleth in its Place," Minnesota Academy of Science, *Journal*, 25-26, 372-384.

RHIND, D. W. (1971) "Automatic Contouring — An Empirical Evaluation of Some Differing Techniques," *Cartographic Journal*, 8, 145-158.

SCHMID, CALVIN F. and EARLE H. MacCANNELL (1955) "Basic Problems, Techniques and Theory of Isopleth Mapping," American Statistical Association, *Journal*, 50, 220-239.

SCRIPTER, MORTON W. (1970) "Nested Means Classes for Statistical Maps," Association of American Geographers, *Annals*, 60, 385-393.

SHEPARD, DONALD (1968) "A Two-Dimensional Interpolation Function for Irregularly-Spaced Data," Association of Computing Machinery, 23rd National Conference, *Proceedings*, 517-524.

Statistics Canada (1972) *GRDSR: Facts by Small Areas*, Ottawa, Canada.

YOELI, PINHAS (1977) "Computer Executed Interpolation of Contours into Arrays of Randomly Distributed Height-Points," *Cartographic Journal*, 14, 103-108.

Computer Cartography as It Is Taught at Louisiana State University

by Philip B. Larimore

THE PAST

Louisiana State University entered the world of computers in 1959 with the installation of an IBM 650 computer. Since that time computing needs have continued to grow, and recent installation of a new IBM 3033 computer will keep pace with the demand for at least a few years.

In 1971, the School of Geoscience (Geography, Anthropology and Geology) recognized the need to have students trained, not only in computer programming, but also in computer graphics. A Milgo DPS-7 flatbed plotter was installed which could handle large work maps (40" by 60"), utilize many different drafting mediums, and plot on preprinted maps. Later that year, it was decided that the School of Geoscience generated sufficient work load to support a full-scale RJE batch terminal. Installed in late 1971, the terminal included a Documation M-600L card reader, a Data Printer Corporation V-132 650 lpm printer (with overprint capabilities and the 6 or 8 lpi option), a Mod 10 Wang magnetic tape transport having both read/write capabilities, a Nova super 1200 minicomputer with 32K memory, an 029 card punch, and a Beehive CRT which is the controller. Later, a Benson-Lehner LARR digitizer was added.

A major feature of the terminal was a policy of "user operation." The work load was such that, even with the employment of a full-time programmer, all persons with input for the terminal had to be trained to operate the equipment. Thus the "hands on" policy was developed. Today all persons using the School's computer room are taught to handle their own jobs. The programmer is usually available if needed, but both students and faculty do everything from reading in to print out or plotting on their own. This has led to some small damage, but most of the time it has worked quite well, and people are free to use the terminal when they need it or have the time.

With the installation of proper equipment in 1971, the Department of Geography and Anthropology instituted a course in computer cartography the following year. Since there was no cadre of computer-trained students, the course was designed to train students to produce good usable computer-generated graphics with little prior computing or cartographic knowledge. There were, at that time, a few canned mapping programs available on the market. Called "cookbook" programs, they could be used with minimal programming knowledge. SYMAP is one of the more outstanding and best-known examples of this type of program. Since that time, numerous programs have become available.

Having no prerequisites and with the course available University-wide, there has never been any problem filling the class. Students and faculty from Landscape Architecture, Engineering, Agricultural Engineering, History, Environmental Design, Urban Planning, Mathematics, as well as Geographers, Geologists and Anthropologists have taken the course. Many students have to be taught to key punch and how to prepare a job card, but they are given the JCL necessary to run any program used in class. Most of the maps used in the class assignments have the outline digitized, not to make the course easier, but to speed up the assignments and permit more time for learning basics.

In the ten times this course has been taught, this system has worked quite well. Computer Cartography is not an "easy" class; students spend many hours in the successful completion of an assignment. But, despite such demands, the class fills each semester and some students have to be turned away.

THE PRESENT

At present Computer Cartography (Geography 4043) is offered in both the spring and fall semesters. The class meets twice a week for an hour and a half per meeting. During this time, a program is introduced, explained, and one or more simple class projects is assigned to be finished by the next class period. As in most classes, the three class hours are simply the tip of the iceberg. Each student spends many hours key punching a job, reading it into the computer, waiting for the job to be run and printed out, correcting errors, and resubmitting it. Each student averages three to four turnarounds per class project. Additional instruction is done during this time, and I average four hours per week per student in individual instruction.

The first program assignment is SYMAP. I have found it to be the best program for beginning students. The manual is well written, easily understood, and results are quickly obtained. Before introducing choropleth mapping, a lecture session is spent discussing the problems of interval selection and several techniques used to calculate class intervals. References and reading assignments on the subject are suggested. It is important that students understand the problems involved, the importance of class intervals, and the number of classes for proper map production. This leads to discussion of types of map users and of the readers' understanding of cartography and graphic presentation. It is my feeling, I think well justified, that most producers of all types of graphics overestimate the level of knowledge and ability of the reader. This is certainly true when preparing maps for the general public.

Four weeks (eight lecture sessions) are spent on SYMAP. Admittedly, this is considerable time in a course that lasts only seventeen weeks, but I feel that it is necessary. During the lecture sessions, the problems of data manipulation, cartographic production, graphic reproduction, and cost factors in computer production are discussed. At the same time the student is working and producing maps.

From SYMAP the next program introduced is SMYVU. By building on the knowledge gained in SYMAP, this step uses much less class time, and in one week the student is producing block diagrams. However, starting this fall ASPEX will replace SYMVU: Following SYMVU (or ASPEX), CALFORM is introduced. With this program the student is introduced to the plotter and is taught to bring jobs in on tape and set up the plotter. After completing the section on CALFORM, POLYVRT follows, and with these four programs (about mid-semester) the student has a good introduction to computer mapping and sufficient skill to see how the programs and their newly acquired knowledge can be used in their own fields. At this point, the class is not expected to understand the programs, only how to use them. In fact, most of the students will never understand many of the programs.

During the second half of the semester, the students continue to produce maps using the four programs introduced earlier, and learn other small programs that have been written by advanced students in the School of Geoscience: CLIMATAGRAPH, a program to plot rainfall and temperature as a curve on a bar graph for any period of time; CIRPLT, a program to plot scaled pie-graphs on a base map; BARGRAPH, a program to draw up to 12 bars on a graph with up to 10 patterns per bar; and SHADO, a program to describe the shadow of any building or group of buildings anywhere on the earth at any time. With students from a variety of fields, it is unlikely that they will use all such programs, but I feel it important to introduce a wide variety of available programs.

Subroutine MAP, a program for drawing projections, follows. This program, written by John O. Ward at the National Oceanographic Data Center, Miami, Fla., has seventeen projections and is written in modular form to allow insertion of data. As a Cartographer, I like this program because it plots outline maps with the parallels and meridians as well as the landmass, and I feel that this presents a map that is more easily understood. The longitude and latitude grid makes area location easier.

At this point I take time to explain the advantages and disadvantages of some of the better known and more widely used projections. Of course, we don't go into detail, but sufficient information is given to enable students to understand which projections are conformal, which are equivalent, which have neither attribute, and what these differences can mean on the maps they produce.

The last major program is SURFACE II. This program, a software system written by R. J. Sampson, is available from the Kansas Geological Survey, Lawrence, Kansas. The basic form produced is a plotter-drawn contour map. It also produces perspective block diagrams, perspective block diagrams for stereoscopic viewing, trend surface and residual maps, and cross sections.

The final two weeks of the semester are spent with each student working on a project of his own selection and preparing an oral class report. The student must make use of as many of the presented programs as possible. This report is not expected to be on the level of a seminar paper or a small thesis; the idea is to have each student see how each program might work in his field using his data.

The primary goal of the course is to produce students capable of creating a variety of computer generated graphics. Although few emerge as experts, all who pass the course can produce results — in itself a marketable skill. A side benefit is that many are tantalized by the possibilities and seek further training. Some come to traditional cartography, while others seek courses in programming and statistics. Many graduate students are encouraged to experiment with these graphics, and several have used this knowledge in their research.

THE FUTURE

The demand and need for training in computer generated graphics at LSU is growing and will continue to grow at a rapid rate. Some major problems remain, but they are being removed as rapidly as funds become available. The recent installation of the IBM 3033 will be a big help by reducing turnaround time, which has been as much as twenty-four hours during peak use periods.

Even with the installation of a new computer, the lack of funds for additional equipment remains a major deterrent to more rapid growth in the School of Geoscience. We have need for a new and better digitizing system, an interactive CRT, a diskpack, a faster plotter, and an additional tape deck. Most of the existing equipment is over seven years old, quite ancient in a field of rapidly changing technology. Even with problems, the demand for class space continues to grow, and the use of our facilities steadily increases. The publication of research papers and atlases by faculty, students, and staff is beginning to attract attention throughout the state. As these papers reach the desks of government offices and private research organizations, where knowledge of computer mapping is not universal, we begin to receive inquiries. How were the maps prepared, could we do maps and graphics for them, or could they send their people to us to learn the process?

Students in many fields realize that a knowledge of computer graphics places them in a better bargaining position when they enter the job market. Planning organizations are making more use of computer mapping and are looking for graduates with this knowledge. Production people are searching for ways to produce graphic material faster, and computer generated atlases can satisfy this demand at lower production costs. As an example, the Louisiana Census Data Atlas, produced using only SYMAP, included more than three hundred maps, yet it took less than six months to produce at a cost of about five dollars per atlas. Admittedly, the maps are not as "beautiful" as hand-drawn maps, but it helped to get a mass of census data into a usable graphic form and into the hands of those who needed it. It greatly reduced the time between data collection and data presentation, lowered production costs, and has been well received by those using it.

The next "logical" step at LSU would be to offer an advanced course, in which the students would be required to know programming. In this course students would write their own programs to solve specific graphic problems. However, with my present duties as teacher, cartographic laboratory supervisor, and supervisor of the computer graphics section, a more realistic step is to try to find sufficient time to offer the basic course for a third time during the school year. We have under consideration a proposal to offer the course at night in the University College, since there have been a number of requests to have it available at a time when business, industry, and government people could enroll. If this proposal is accepted, as the punch line in an old joke says, "there goes Sundays."

Computer Graphics in the Interpretation of Cell Kinetics

by Thomas L. Lincoln, M.D., and Stanley E. Shackney, M.D.

My major interest in computer technology is as a user; I use computer graphics as an interpretive tool in cancer research. In this paper, I will discuss what we are doing now, what we would like to do, and some of the data manipulation requirements which underlie our use of graphic displays.

In one research application, we are studying how leukemia and lymphoma cancer cells progress through the cell replication cycle from cell division to cell division and, as a consequence, how tumors made up of these cells grow. The purpose is to characterize tumor growth in man and to interrupt tumor cell proliferation using chemotherapeutic drugs. At the same time, normally dividing cells must be spared. Graphic displays assist in interpreting the underlying biological data.

This field, which is called cell kinetics, has a 20-year modern history. In the early 1960's, Puck (see footnotes 1 and 1a), working in tissue culture, showed that cells prepare to divide by going through a sequential ritual. Following division, cells may spend a certain amount of time in a dormant state; then they enter a phase of active metabolic preparation; next they double their genetic material and copy its code; then they spend some time in error correction; and then they divide again. In normal cells the cellular material, including the genetic material, is divided equally between the daughter cells. Some cells take a separate path, grow up, and stop dividing. All along the way, some cells die.

These basic observations, once made by laborious techniques, remain the foundation of cell kinetics. The same basic sequential behavior is characteristic of both replicating normal cells and replicating tumor cells. Differences between these cell types are important, and lead to differences in the effect of chemotherapy. Tumor cells pass through the cell cycle sequence at different rates from normal cells; the genetic material is sometimes unequally divided; and cell death is sometimes more pronounced. All of these features underscore the observation that tumor cells have lost some of the orchestrated set of controls which govern normal cells. Ultimately we would like to know what controls have been lost. For now, we measure these differences by measuring the presence and the quantity of cell constituents which are related to the cell cycle.

Present technology provides us with automated equipment which will allow multiple measurements of cell constituents on millions of individual cells in normal or tumor samples. One set of such measurement devices is the flow cytometers. Each cell is measured as it flows in a capillary-size droplet or fluid stream using laser probes and/or light scattering techniques. Today, as many as four measurements can be made on each cell. These measurements stand surrogate for the growth characteristics which we would like to know. For example, we can measure DNA content, RNA content, enzyme concentrations, cell size, nuclear size, and certain changes in physical state (such as hydration), each of which can be related to some aspect of the replication cycle. With each new device, more measurements are possible. When flow cytometer measurements are made, we obtain a snapshot — a static picture of a dynamic growth process. The most reproducible measurements are made under experimental conditions on pure cell populations in steady state growth. These experimental systems form the basis for cell kinetic models and also for the interpretation of more complicated clinical material.

To introduce our use of graphic displays, Figure 1 is an example of the frequency distribution of cell size and DNA content in an actively proliferating population of normal white cells from the blood. This display, and the displays which follow, are programmed at Los Alamos as a part of the flow cytometer instrumentation developed under an interagency agreement with the National Institute of Health. The measurements on lymphomas are part of a collaborative effort between Stanley Shackney, M.D. at the National Cancer Institute and the Southern California Lymphoma Group under the leadership of Robert J. Lukes, M.D. In our field, these displays are state-of-the-art.

The similarity of Figure 1 to those produced by three dimensional mapping programs is evident. The terrain is an abstract one. The simultaneous measurement of cell volumes and DNA content defines the plane. The frequency of cells with these characteristics defines the elevation. Some smoothing is required. When a calibrated focal point is combined with a conventional topographic orientation, the display can be considered in one sense or another as a map. Thus, the conventions of cartography provide a means of displaying these data so that multiple relationships among the variables can be perceived and interpreted. In this normal example, most of the cells are in a single peak.

Note that most of the blood cells are small with an unduplicated DNA content. As the cells increase their DNA content and duplicate their genetic information, they also increase in volume so that the replicating cell population falls along a diagonal in the horizontal plane. The diagonal represents the cell cycle sequence from division through DNA synthesis to division again.

The separate two-dimensional projections presented in Figures 2 and 3 are less insightful than the three-dimensional presentation. An isometric display of these same data (Figure 4) shows the fundamental thematic relationship of replicative growth and cell size. Here the diagonal relationship representing increasing cell volume is very evident. The display also

[1]Puck, T.T., and J. Steffen, "Life cycle analysis of mammalian cells. I. A method for localising metabolic events within the life cycle, and its application to the action of colcemide and sublethal doses of X-irradiation," *Biophys. J.*, 3, 1963, pp. 379-397.

[1a]Puck, T.T., P. Sanders, and D. Peterson, "Life cycle analysis of mammalian cells. II. Cells from the Chinese hamster ovary grown in suspension culture," *Biophys. J.*, 4, 1964, pp. 441-450.

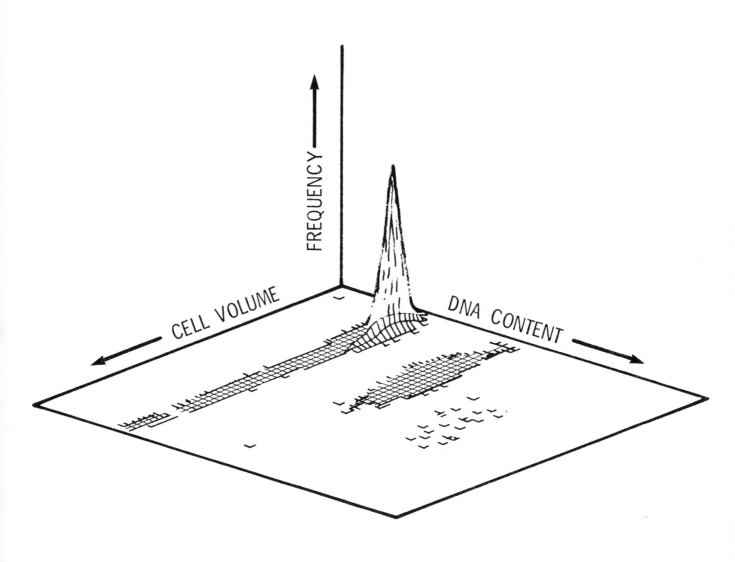

Figure 1. Normal White Cell Population

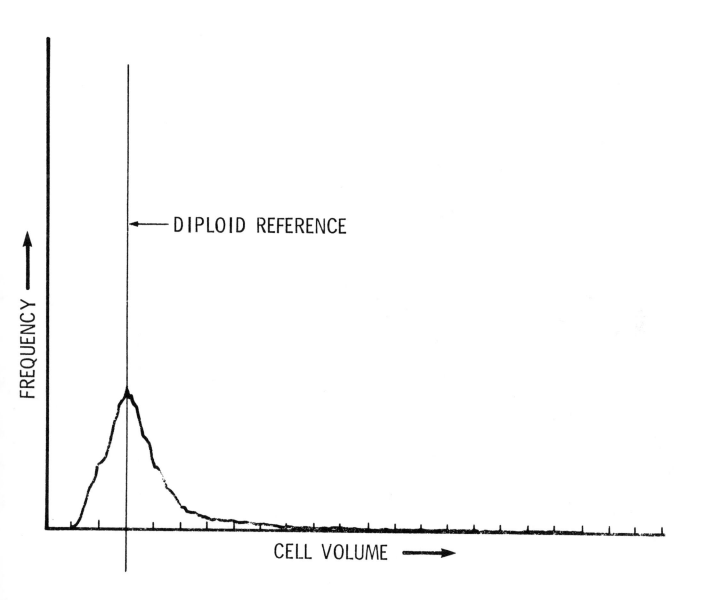

Figure 2. Distribution of Cell Volume for Large Non-Cleaved Lymphoma

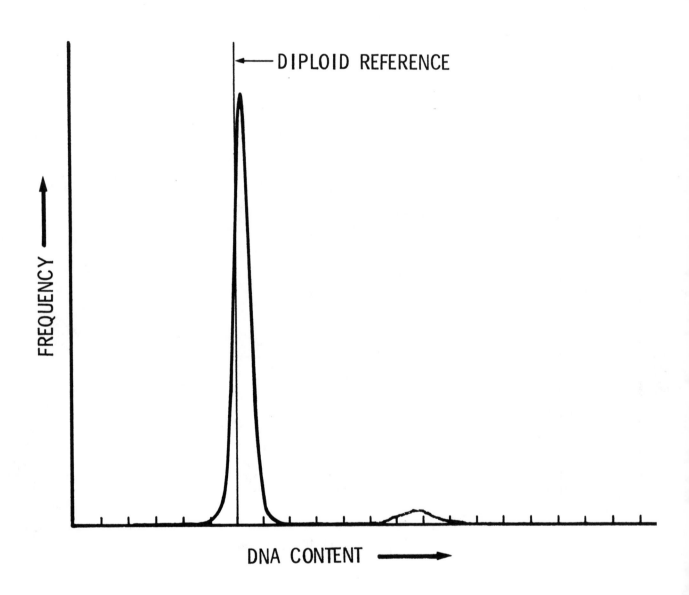

Figure 3. Distribution of DNA Content for Large Non-Cleaved Lymphoma

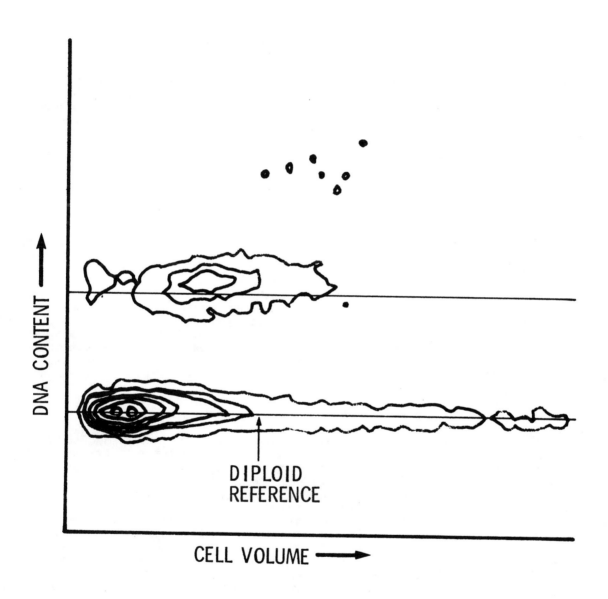

Figure 4. Isometric Display of DNA Content vs. Cell Volume

shows the two-dimensional calibration of resting-cell DNA genetic content and resting-cell size.

Although we now understand the components of the cell replicative process more clearly than we did ten years ago, the biological characterization of the cycle remains incomplete. There is more detail and variety than we originally imagined. The population of cells which we observe do not pass through the phases of the cell cycle at exactly the same rate, but rather progress with some time distribution. Although it would lead us far afield to consider the details,[2]and[3] the interpretation of these measurements can be formalized in distinct and conflicting mathematical models, all of which can be made to fit the basic data for a steady state growth. One example of such a model, proposed by our group, is given in Figure 5.[3] Here the progression through the replicative cycle is represented as the horizontal axis and the rate of progression is represented by the axis in perspective. The vertical axis is a normalized density function. The density distribution reflects a population which is in exponential growth. Other models could interpret the same data in other ways. Significant differences in model predictions can be observed when cell populations are perturbed. These distinctions, when compared to data, should verify the correctness of one model over another. However, the experimental perturbations introduce new uncertainties and ambiguities.

Graphics allows us to search out consistent patterns of biological behavior in our multivariate data, and to observe the graphic expression of a theme without too much initial concern for mathematical or modeling details. The ability to view multiple variables simultaneously provides a context for interpretation based on such consistency. Thus, computer graphics gives us a better means of "eye-balling" the data as we work toward a deeper analytic understanding.

This capacity to abstract a theme gains further importance as the problem of interpretation becomes more complex. In our laboratory, we are primarily concerned with clinical material which has been obtained from patients in the course of their medical diagnosis and treatment. This material contains mixed cell populations where some cells are normal and some are abnormal. Repeatable laboratory conditions do not exist. Moreover, the tumor is not a standard laboratory tumor and has its own variability and unknown characteristics.

In Figure 6 we present another abstract terrain. The example is a lymphoma growing in the bone marrow. There are peaks and valleys in the distribution of cell size in this space. The two highest peaks distinguish the normal and the abnormal cells because the lymphoma cells are markedly larger in volume. There is a population of smaller normal cells. Each cell type is undergoing replication, copying their genetic memory which is contained in the DNA-coded material. Each cell type is increasing in volume as the replicative process progresses so that the cell cycle appears to follow the diagonal. Again, using conventions borrowed from cartography, isometric displays are helpful (Figure 7). Different tumor types present different patterns (Figure 8), which allows us to interpret the underlying biology (Figure 9).

In our examples, there are a number of distortions and artifacts which are a consequence of our measurement techniques. These are cell volume distortions which give the long

[2]Steel, G.G., *Growth Kinetics of Tumors*, 1977, Oxford University Press, London, England.

[3]Lincoln, T., P. Morrison, J. Aroesty, and G. Carter, "Computer Simulation of Leukemia Therapy: Combined Pharmacokinetics, Intracellular Enzyme Kinetics, and Cell Kinetics of the Treatment of L1210 Leukemia by Cytosine Arabinoside," *Cancer Treatment Reports*, Vol. 60, No. 12, December 1976, pp. 1723-1739.

tail to the volume distribution. There are cells which are stuck together and thus measured together, giving artificial multiples of size and of DNA content. These can become confused with the measurements of single cells. Thus it becomes important to judge which components of the display can be discounted as artifacts. A major purpose of the graphic display is to aid in this discounting procedure. As a first approximation, there is a visual cropping of the generated display, for example, removing the tails noted above. Another simple example is the separation of background noise from signal. The basic plane for measurement must be chosen in such a manner that the biological pattern becomes clear. At the same time, one must guard against a choice which merely gives weight to preconceptions. As in all pattern recognition, we come up against the problem of how much *is* and how much *should be* in the eye of the beholder. We need to look at the now data and add stylized conventions later.

In diagnostic medicine, the ultimate test of a procedure is predictability. The measurement snapshots which we generate and the interpretations that we give them should so reflect the underlying biology of cell growth that a prognosis about a particular patient can be extrapolated into the future.

These are the displays which we can create now, but more complicated displays would be helpful to give us more confidence in the process of thematic abstraction. A further dimension could be added by introducing color on a red to blue scale — for example, a measure of RNA content as a surrogate for protein synthetic activity. It is not beyond consideration to add "texture" or shading as a further variable — perhaps to highlight the presence of a particular enzyme, or to introduce degrees of brightness — particularly on a CRT — to provide, for example, a semi-quantitative description of variance. In this fashion we could visually integrate five or six separate variables in an abstract map. If such displays could not only be generated but also manipulated using computer commands, then the implications in these data could be explored interactively and unanticipated questions could be asked. In appropriate circumstances, where sampling can be done as a function of time, we could watch the cinematographic evolution of such maps — a variation on computer animation. This would bring us very near our ultimate objective — to understand the transient behavior of a tumor system which has been perturbed by chemotherapeutic drugs.

The capacity of appropriate mathematical models to generate an equivalent graphic sequence could test the adequacy of our abstractions. The techniques which I have described represent an integrative approach to data. For some who are analytically inclined, it might seem to be moving in exactly the wrong direction — the five or six tables of data which we can extract from our measurements might appear to be ends in themselves or ready for algebraic interpretation using regressions or other techniques. However, for pathologists, who tend to have a geometric turn of mind, the construction of an interpretable context as a means of dealing with fuzziness and ambiguity is a more comfortable approach. We are accustomed to search for consistency based on a visually weighted perception of events. This is what one does with a microscope.

I have presented computer-based cartography as a tool for identifying patterns in multivariate biological data. However, the broader issue is not merely the display of data but the ease of on-line manipulation of data which has been organized in tables.

What generally holds us up now as biological researchers and computer users is the reorganization of our data once generated and initially tabulated. If this prosaic process could

$$\frac{\partial n}{\partial t}(t,a,T) + \frac{\partial n}{\partial a}(t,a,T) = -\lambda n(t,a,T)$$

WHERE $N(t) = \iint n(t,a,T)\, da\, dT$

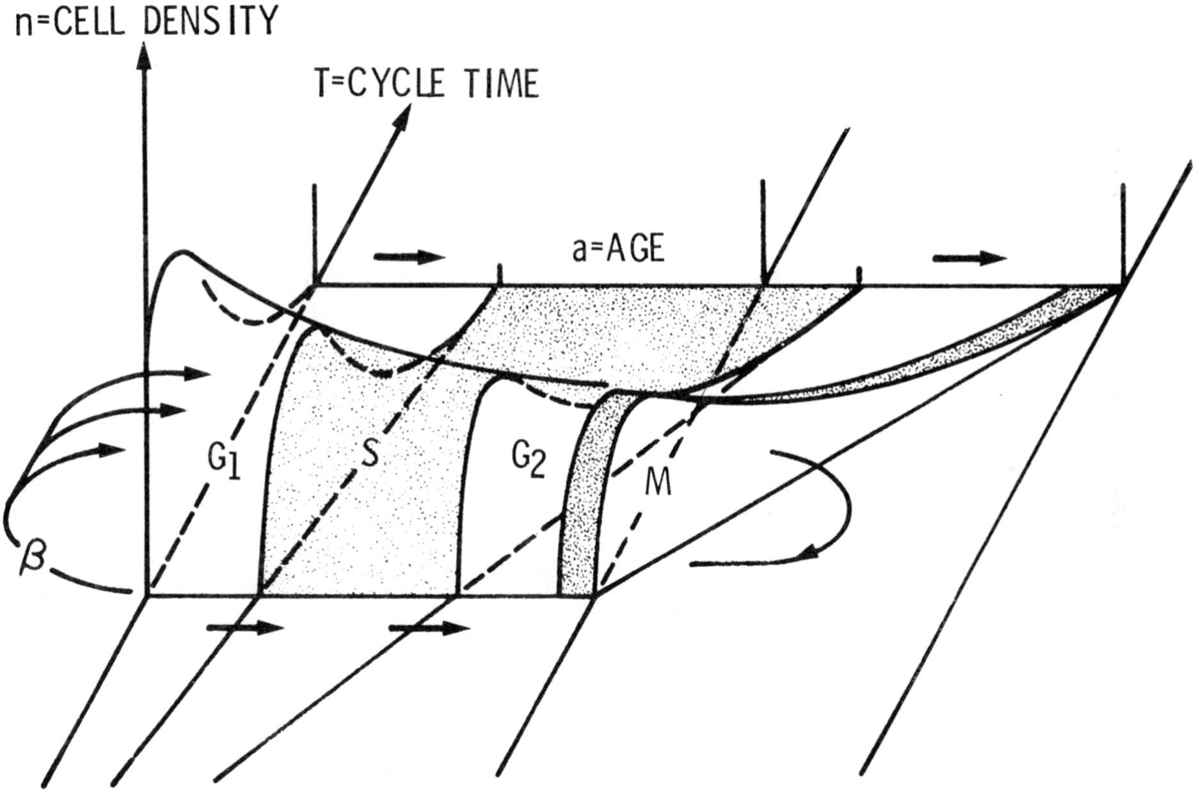

n=CELL DENSITY

BIRTH PROCESS

$$n(t,0,T) = \beta \int K(T,T')\, n(t,T',T')\, dT'$$

WHERE $K(T,T') = G(T)$

$G(T)$ = DISPLACED GAMMA DISTRIBUTION

Figure 5. Cell Kinetic Model

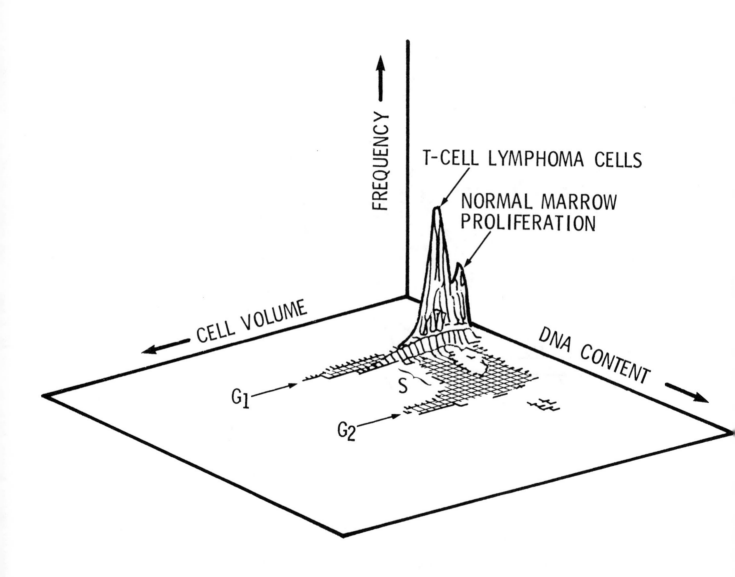

Figure 6. Flow Cytofluorometric Measurements on Bone Marrow in a T-Cell Lymphoma Leukemia Distinguishing Normal Cell Kinetics from Tumor Cell Kinetics

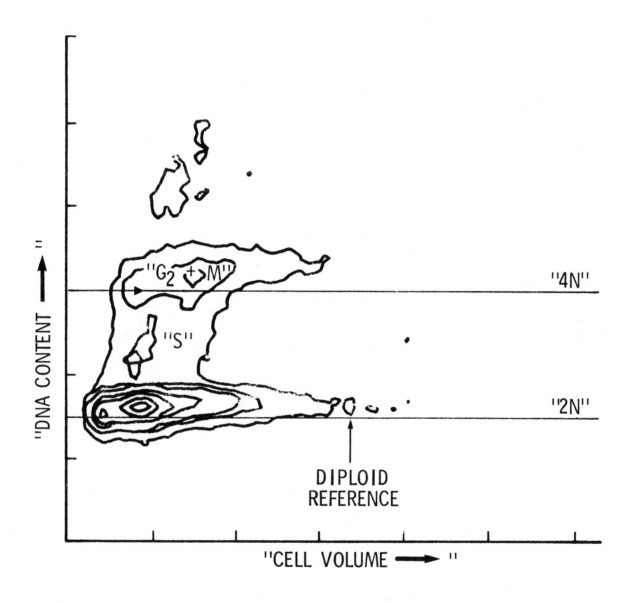

Figure 7. Isometric Display of Bone Marrow From T-Cell Lymphoma-Leukemia

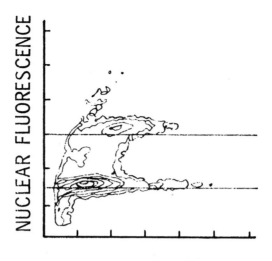

a. Immunoblastic Sarcoma (Histiocytic Lymphoma)

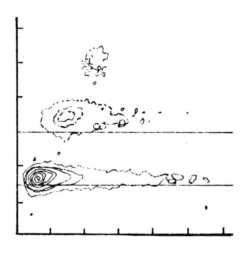

b. Small Cleaved (Poorly Differentiated) Lymphoma

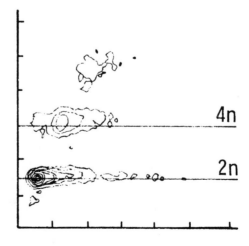

c. Chronic Lymphocytic Leukemia

Figure 8. Coulter Volume

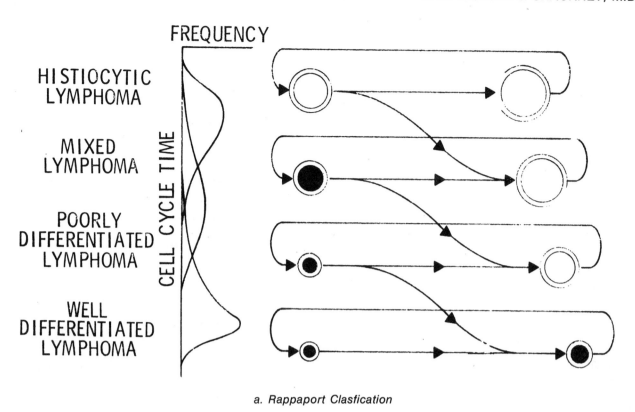

FREQUENCY

HISTIOCYTIC
LYMPHOMA

MIXED
LYMPHOMA

POORLY
DIFFERENTIATED
LYMPHOMA

WELL
DIFFERENTIATED
LYMPHOMA

CELL CYCLE TIME

a. Rappaport Clasfication

IMMUNOBLASTIC
SARCOMA

FOLLICULAR CENTER
CELL LYMPHOMAS:
LARGE
NON-CLEAVED

SMALL
NON-CLEAVED

LARGE
CLEAVED

SMALL
CLEAVED

SMALL LYMPHOCYTIC
LYMPHOMA (CLL)

FREQUENCY

0   60 120
DNA CONTENT

FREQUENCY

0   60 120
DNA CONTENT

b. Lukes Classification

Figure 9. Non-Hodgkin's Lymphoma (B Cell)

be accomplished in a more orderly and convenient manner, we could make better use of the graphics that are available. Our use of graphics and the proposed extensions are specific and somewhat esoteric examples of a very general need for what might be called "data table processing" — the interactive reorganization of data organized in tables to answer questions which were not initially anticipated.

In order to explore multivariate data heuristically, the user may wish to call upon a vocabulary of commands which interrogate the underlying table structure and generate secondary tables. Cartography, and the manipulative and display conventions of cartography, represents one powerful way of presenting multivariate data, although the underlying data need not be related to maps. On the other hand, the user may wish to examine the same data using algebraic statistical techniques, or merely list his data with a different juxtaposition or in a different sequence. In this scenario, the user calls upon a set of table-oriented utilities, by means of simple commands, to restructure his data. He operates in the same manner as he would using a text editor — modifying a scratch copy of his data to meet an evolving set of questions. One might suppose that data table processing, like text editing, would be widely available and widely used. However, this is clearly not the case. Most command languages constrain the user to remain within a narrowly predefined vocabulary. Within the specified design range, the system may be very powerful, but simple utilities which connect the system at hand with the messy world of data tables in general are commonly missing.

There is a long list of programs which are powerful, but parochial in the sense described. Any major computer center has a set of such packages of which they are justifiably proud. STATLIB,[4] a powerful statistical package jointly developed by Rand and Bell Labs, is one example. BIOMOD [3]and[5] a graphic biological modeling language developed at Rand, is another.

Many interactive graphics packages can be manipulated through a full range of orientations and magnifications which dramatically demonstrate their potential computational proficiency and virtuosity. But the data table is generated by the graphic designer and is conveniently distant from the problems of raw data input. Cartography programs which incorporate real world data can in part redress this balance. Moreover, by providing visible examples of data manipulations, cartography can give a potential user an intuitive grasp of the power of data table processing.

Philosophically, CLINFO [6]and[7] is a tentative step beyond the limits of these applications. It is a set of programs developed at Rand under contract for the Division of Research Resources, NIH, to deal with the storage, retrieval, formatting, reformatting, graphic display, and statistical processing of time-dependent clinical and experimental data. This package represents a primitive but powerful tool for defining and working with data tables. If we could strengthen this core and make data table processing more flexible and convenient, the power of graphics would come to its full fruition.

[4]Brelsford, W.M., and D.R. Relles, "Interactive Statistical Computing, with Applications to Forecasting and Data Analysis," *Proceedings of Computer Science and Statistics: 8th Annual Symposium on the Interface*, 1975, Western Periodicals Company, Los Angeles, California, pp. 530-533.

[5]Groner, G., and R. Clark, "BIOMOD: An interactive graphics system for analysis through simulation," *Miami Beach, Florida: Proceedings of the 1971 IEEE Conference on Decision and Control*, pp. 197-201.

[6]Groner, G.F., M.D. Hopwood, N.A. Palley, and W.L. Sibley, *An Introduction to the CLINFO Prototype Data Management and Analysis System*, The Rand Corporation, R-1541-NIH, December 1977.

[7]Hopwood, M.D., G.F. Groner, N.A. Palley, W.L.Sibley, et al., *An Evaluation of the CLINFO Data Management and Analysis System*, The Rand Corporation, R-2260-NIH, November 1977.

The Use of CAMS (Computer Aided Mapping System) as a Teaching Tool

by Robin S. Liggett

INTRODUCTION

Computer mapping techniques have numerous applications in the areas of architecture and urban planning. For a number of years, standard mapping packages have been available as a tool for faculty and students at the School of Architecture and Urban Planning at UCLA. Typical applications of these packages include two-dimensional mapping of census data for demographic studies and three-dimensional contour-map generation for architectural sites.

There has been a reluctance on the part of students and faculty with little or no computer experience to make use of the available mapping packages. Data gathering and punched deck preparation were time-consuming and job control language tedious. Turnaround for producing maps was not normally within the time frame of a class project. A key problem in data preparation concerned the specification of geographic coordinates for a map. If prepared manually, this phase was too time-consuming to be reasonable except for large projects where many maps were to be produced.

Over the past few years, a concentrated effort has been made to increase the use of the computer facilities available at the school. Faculty and students are encouraged to think of the computer as an easy-to-use and accessible aid for a variety of class and studio projects. For this to be possible, the use of the computer must be simple and cost effective. To be cost effective, computer generated output must either be of significantly higher quality than results which could be obtained manually, or the generation process must be less time-consuming.

The acquisition of computer graphic equipment by the School of Architecture and Urban Planning (specifically a graphics tablet which could be used for digitizing locational data) made the development of computer systems that would address the above concerns feasible. One of the first projects undertaken was the development of a Computer Aided Mapping System (CAMS). CAMS is an interactive system for computer mapping and census retrieval which incorporates well-known computer mapping packages. It provides the user with an interactive framework for specifying the input data for these packages and initiating batch jobs to run the packages. The user need not be familiar with job control language or specific data formats of the individual packages, rather he need only answer sets of programmed questions.

The key feature of CAMS is the incorporation of existing software within a framework which facilitates use by students and faculty without prior computer experience. Although standard packages still run in batch mode, the user prepares all data and submits jobs in an interactive mode. CAMS provides a structure for communication between a set of packages which produces two- and three-dimensional character and plotter maps. Components developed at SAUP include interactive in-terfaces to standard packages as well as special programs for digitizing geographic data and retrieving census data.

The educational uses of CAMS are twofold. First, the development of the system was undertaken by an advanced computer applications class consisting of graduate planning and architecture students. Many of the components of the system were developed as student projects. Second, a major goal of the system was to provide a medium for introducing the use of the computer into planning and architecture classes. This paper will present the specific features of CAMS, its classroom applications, and a variety of student projects which illustrate the use of the system.

CAMS

The School of Architecture and Urban Planning Computer Laboratory currently houses two TEKTRONIX 4012 graphics terminals, a SUMMAGRAPHICS 48-inch digitizer, and a VERSATEK 1200A printer/plotter. This equipment accesses, via TSO, the campus mainframe (an IBM 360/91) over two 1200-baud direct lines. The basic hardware requirement of the CAMS system consists of one TEKTRONIX terminal, the digitizer and a printer/plotter.

CAMS software can be categorized into two classifications: standard mapping packages obtained from outside sources and interactive data preparation programs developed in-house.

Programs obtained from outside sources are: SYMAP: for producing two-dimensional character maps which graphically depict spatially disposed quantitative and qualitative information (Harvard University)[1]; SYMVU: for generating three-dimensional surface plots on CALCOMP or VERSATEK plotters (Harvard University)[2]; and CELNDX: for manipulating spatial variables stored in regular gridded structures and providing "overlay" mapping capabilities (Oakridge National Laboratory)[3].

Each of these standard mapping packages has an interactive interface, developed by SAUP, which allows the user to specify key parameters needed for execution. Programs in this category include EASYMAP, interactive SYMAP control card specification; EASYVU, interactive SYMVU control card specification; and OVERLAY, interactive CELNDX control card specification.

The CAMS system also provides a set of interactive programs for data preparation. This data is stored in files easily accessible to the mapping packages. Programs in this category

[1] **SYMAP**, Laboratory for Computer Graphics and Spatial Analysis, Harvard University, Cambridge, Massachusetts.
[2] **SYMVU**, Laboratory for Computer Graphics and Spatial Analysis, Harvard University, Cambridge, Massachusetts.
[3] Wilson, D.L., **CELNDX, A Computer Program to Compute Cell Indices**, Oakridge National Laboratory, Oakridge, Tennessee, 1976.

include INTAB, interactive coordinate or outline specification (via a graphics tablet); RETCEN, census data retrieval; and TOTCEN, subtotals of district or zonal census data.

For each of the interactive programs in the CAMS system, separate user documentation is available. This documentation is in the form of 3-to-6 page "write-ups" for each program module, and assumes no prior user experience with either mapping/retrieval packages or computers. Each write-up includes a brief discussion of the capabilities and objectives of the program. A step-by-step explanation is provided in each write-up which expands and clarifies the messages issued by the program. These write-ups are kept in an online data set which allows multiple copy generation on demand and facilitates updates and corrections. Throughout the documentation process, attention has been given to standardization of the write-ups.

Figure 1 shows the entire system in modular form. The specific modules chosen by the user will depend on the type of map to be produced. Possible paths through the system are illustrated with specific examples in the next section.

CLASSROOM APPLICATIONS

Computer mapping using CAMS is presented as a special topic in the introductory computer course for architects and planners. This introduction is intended to acquaint students with the computer resources available and encourage their use in non-computer classes. Extensive documentation also makes CAMS readily accessible to all students and faculty.

CAMS has been used for a variety of applications which are representative of the many different areas of study available at the school. These include architecture and urban design, environmental planning and analysis, delivery of public service systems (transportation, education, health care), social policy research, and urban/regional development. The following examples have been selected to demonstrate how CAMS can be used in a wide range of applications.

Example 1: Contour Mapping (Site Elevations)

One path through CAMS allows the user to produce two- and three-dimensional contour maps (e.g., depicting the elevations of a site). Using the INTAB program and a topographic type map, contour lines are digitized (i.e., geographic coordinates associated with a line are entered into the computer) and corresponding values (elevations) are recorded. This data provides the required input which the SYMAP program needs to produce a two-dimensional character map. The EASYMAP program is run interactively, allowing the user to specify the required SYMAP control parameters and dataset names. When information input is complete, EASYMAP submits a batch job to create the map on the line printer.

After a satisfactory 2-D map from SYMAP has been obtained, the user may execute the EASYVU program in order to specify SYMVU control parameters and produce a 3-dimensional plot.

Figure 1 exhibits a 2-dimensional character map and a 3-dimensional pen plot generated for a contour mapping application. This project, executed for a landscape design course, enabled the architect to visualize the terrain of the proposed site. A large computer-generated perspective was used as the basis for the final presentation of the completed design.

Example 2: Census Mapping (Los Angeles County)

Geographic coordinates are available on file for easy mapping of 1970 census data by tracts for Los Angeles County. To map the entire county, the user simply uses RETCEN (the census retrieval module) to retrieve the desired data, and then executes the EASYMAP program selecting the L.A. census mapping option.

Two-dimensional character map output was used in a study concerned with transportation for the elderly.[4] Specifically focusing on Los Angeles County, the study investigated spatial variation in the travel demands of the elderly. A set of computer maps was generated to document characteristics of the elderly which were particularly relevant to transportation demands and needs. These maps, in conjunction with maps displaying the existing public transportation in Los Angeles, can be used as an aid for future transportation planning.

If a smaller region of the county is to be mapped, the user first outlines the region using the graphics tablet and the INTAB program. This program creates a file of census tract ID numbers and geographic coordinates associated with the outlined region. The RETCEN program may then be used in conjunction with this file to retrieve census data for the outlined region. The census value information from RETCEN and the geographic coordinate information from INTAB provide the required input data files for the EASYMAP program. The user then executes EASYMAP to produce a two-dimensional character map of the specified region.

A similar procedure allows the user to aggregate census tracts into a set of districts by outlining the desired regions. After census tract data is retrieved from RETCEN, district totals can be produced using TOTCEN. Outline coordinates and total values are then input to EASYMAP to produce district maps.

An example of this procedure involves a study of crime and the elderly. For this project, maps were produced to compare census data with crime statistics. Since crime statistics were only available at the police district level, census data was aggregated to conform to these districts. Maps depicting district crime rates and the distribution of the elderly population in the city of Los Angeles provide a quick visual aid for determining if crime is more prevalent in areas with large elderly populations.

Example 3: Overlay Mapping

Overlay mapping combines data stored in gridded form into a single-character map using additive weight techniques. Grid maps which are input to the OVERLAY module can be prepared in a variety of ways using the basic data preparation packages of CAMS and the EASYMAP/SYMAP module. Typical uses of overlay mapping techniques have been environmental analysis and site selection.

One school project involved the investigation of alternative development concepts for Century Ranch, a 2,700 acre, ex-movie ranch in the Santa Monica mountains.[5] One aspect of the study consisted of tabulating the various physical characteristics which areas of the site possessed, and the impact each characteristic might have on the suitability for

[4]Bunker, James B., "Determining the Spatial Variations on the Travel Demands of the Elderly: Development of a Methodology," Master's Thesis, University of California, Los Angeles, 1976.

[5]*Century Ranch: Alternative Development Concepts*, Urban Innovations Group, Los Angeles, California, 1972.

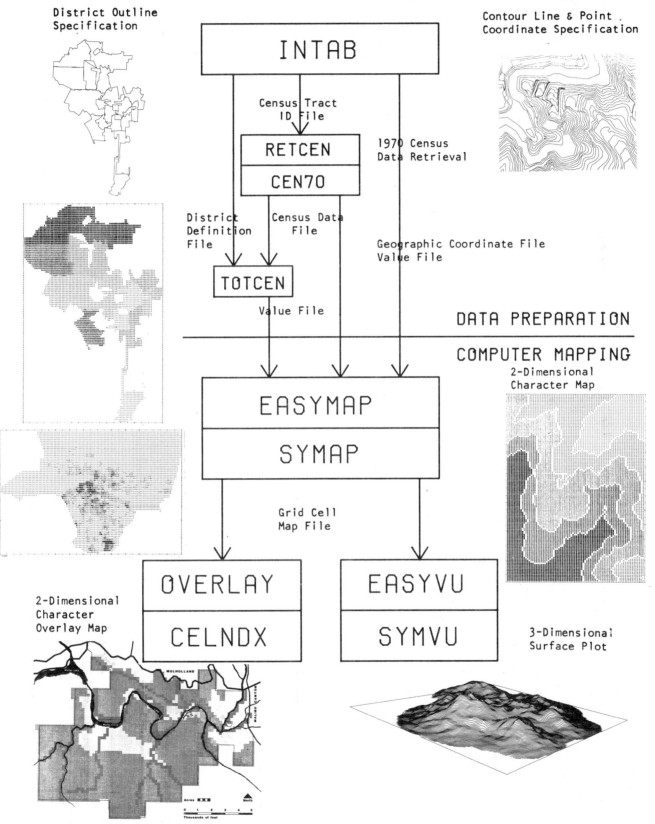

School of Architecture and Urban Planning, UCLA

CAMS: COMPUTER AIDED MAPPING SYSTEM

District Outline
Specification

Contour Line & Point
Coordinate Specification

INTAB

Census Tract
ID File

RETCEN

CEN70

1970 Census
Data Retrieval

District
Definition
File

Census Data
File

Geographic Coordinate File
Value File

TOTCEN

Value File

DATA PREPARATION

COMPUTER MAPPING

2-Dimensional
Character Map

EASYMAP

SYMAP

Grid Cell
Map File

2-Dimensional
Character
Overlay Map

OVERLAY

CELNDX

EASYVU

SYMVU

3-Dimensional
Surface Plot

Figure 1. CAMS: Computer Aided Mapping System

building development. The computer-generated overlay map shown in Figure 1 was produced by combining the individual maps representing this set of characteristics (i.e., topography, geology, hydrology and ground cover), and illustrates the distribution of potentially developable areas of land across the site. Judgements made in weighting the various characteristics could not be completely objective. However, the computer mapping process made it very simple to rerun the data in order to explore the implications which adherence to different criteria may have for use of the site.

CONCLUSION

CAMS has been extremely successful as an educational tool. It has provided a needed link between the student and the computer. CAMS has introduced the student to a variety of methods for data display and analysis which can be used in class or research projects. Again, the primary goal in the design of the system was to simplify the use of existing mapping software by incorporating it in an interactive environment. The number of students currently using CAMS for an increasing variety of applications seems to validate this approach.

PLOTALL: Software Development and Marketing In A University Environment

by Gayle A. Seymour and Rodney S. Marshall

INTRODUCTION

This paper will address three areas of interest to the potential software developer. First, it will touch on the question of software development in an academic environment. Next, it will discuss some of the problems encountered in the marketing, distribution and support of a software package (PLOTALL) from this environment. Finally, an attempt will be made to develop an informal list of considerations which may be useful to academic institutions interested in marketing a package for the first time.

PLOTALL was developed by the Academic Systems/Programming Section at the University of Akron Computer Center. This group, consisting of six programmer/analysts and a manager, is tasked with liaison support for a school of 20,000 students and several hundred faculty members. The university environment is a natural setting for software development. The wide range of disciplines provides strong, objective influences on any developmental venture. Additionally, concentrated testing is readily available in the form of student use. It must be recognized, however, that any software development done in this environment is in addition to the normal support activities. Development of large or complex packages is, therefore, painfully slow and often suffers from changes in direction of commitment.

HISTORY

PLOTALL was designed and written to meet the needs of several departments that were looking for an easy-to-use, generalized plotting package. Initially intended only for "local" use, the project was assigned to one programmer as part of his normal responsibilities. After a system survey was completed, it was determined that a language translation program (which would permit dynamic input) would be most desirable. The language processor approach was also seen as a personal challenge and a learning opportunity for the PLOTALL author.

It was decided to write the language processor in the IBM 360/370 ASSEMBLER language and the Graphics routines in ANSI FORTRAN. This decison allowed for maximum flexibility in writing the language processor, as well as providing easy access to existing and manufacturer-supplied (CALCOMP) graphics software. The initial effort consumed nearly two years of low priority and personal time. The result was a useful, testable, locally derived package which was immediately put to use by instructors in the business college. The package consisted of a flexible language processor which could interpret English-like statements and produce simple line plots, bar plots and pie diagrams. Used in a batch processing environment, the "PLOTALL LANGUAGE" served as an excellent orientation

tool. Its free form, fully edited input allowed for identification of keypunch, syntax and logic errors while still producing basic plots.

The language was used for over two years with only minor changes and maintenance. During this time increased usage produced many requests for improvements and refinement of the graphics routines. With the language processor fully tested, a management decision was made to upgrade the package (graphically) with an eye toward the possibility of "marketing."

A nine-month concentrated effort of graphic and software design and development produced what may now be referred to as PLOTALL VERSION 1.0. This version used the original language processor (written in IBM ASSEMBLER) and a significantly enhanced applications graphic package including new options and improved/optimized graphic routines.

In January of 1978, testing began on the most current edition of PLOTALL. Version 2.0 uses a FORTRAN language processor which significantly reduces the product's machine dependency.

PLOTALL Version 2.1, consisting of several new or improved graphic options, will begin testing in the fall of 1978, and should be ready for distribution by January of 1979.

MARKETING

Once a decision is made to market a package, an entirely new set of problems confronts the software developer. Most university applications software groups are familiar only with the consumer's side of marketing. This perspective can create a false sense of security. That is, it may appear that you are already familiar with most of the problems which you will face. The truth is that the decision to market a package is not one to be made lightly. If the marketing task is to be performed by the software staff ("in-house"), it should be recognized at the beginning that the staff is not trained in the legal or financial aspects of marketing. A learning process will therefore have to take place before an effective marketing effort can be mounted. This learning process is no small task as professionals in the field of marketing would attest to.

Perhaps the point which should be most well defined (but often is not) is: "Why are we trying to market this package?". The rationale for spending the time, energy and money should be determined and reviewed throughout the marketing effort.

The marketing goals developed for PLOTALL included: (1) financial return on time invested, perhaps even a profit (over the long run); (2) good will/good public relations for the University.

Once we had clearly decided our rationale, our first step was to attempt to protect our investment. Through the work of the University of Akron's Legal Staff, the PLOTALL package was copyrighted and the name PLOTALL was given a trademark. A number of questions remain unanswered, however: (1) What is the true value of a copyright for computer software? Courts are presently ruling that computer software cannot be copyrighted. It seems that only the name can be. (2) Once the product has been copyrighted, how and when should potential violations be challenged? (3) How much money will it cost to protect the copyright? It could cost more than the package is worth.

The next order of business was to determine a price for the package. It appeared that our primary market would be other universities. Since PLOTALL had achieved such a high degree of success on our campus, it seemed natural to assume that other universities, particularly our sister state schools, would be interested. Since very few other packages were available for doing similar types of graphics, the price of PLOTALL was set by a general review of other software contracts held with other universities. It appeared that a $500 yearly lease cost to universities would produce some return on our investment and yet would be low enough to entice inquiries. It was also felt that if the package were successful, further development could warrant an increase in the package cost. Price to commercial users was set at $1,000, based also upon precedence set down in other contracts.

Originally, no consideration was given to a purchase option. It was realized at a later date that many commercial customers as well as universities prefer or even require a purchase arrangement. This posed the problem of determining a fair price. It was finally decided that the purchase price should be approximately equal to five years of lease. The subject of purchase also required the need to offer some type of maintenance/update feature (provided automatically under a lease arrangement). This service was made available on a yearly basis at the rate of 10% of the purchase price.

Development of the contract followed much the same line as did the pricing; that is, a complete review of applicable contracts was made and the better points of each were combined into a draft version. This draft was critiqued by the University Legal Staff and a final lease contract was produced. The problem of sale was solved by simply editing the lease contract and substituting "purchase" for "lease." An entirely new contract was not deemed necessary.

This last point has proved to be especially true since each customer seems to desire some new twist or guarantee. Each request is considered on its own merit and all reasonable changes or addenda are allowed.

The PLOTALL user's manual, written by the PLOTALL staff, was published locally by the University Publications Department. Several commercial publication services were considered. Services provided by these companies included editing facilities, additional advertising, and production-oriented features such as warehousing and distribution of manuals. Since our original objective was to provide a quality, low-cost manual for local student use, the use of the "in-house" publication seemed appropriate. Local publication and distribution of manuals has, however, proven to be problematic. The administration and accounting of these manuals for both local and commercial use is simply one additional task for the PLOTALL staff to be concerned with.

With the groundwork laid, the next phase of marketing was advertising. It was decided to carry out a general advertising campaign to more or less test the market. This effort took several directions. First, general news releases were given out by the University News Service. Brochures were prepared by the University Publications Department under the guidance of the PLOTALL "team." Introductory letters were prepared and a

list of potential users (both university and commercial) was compiled from a number of sources.

The mail-out campaign produced a larger number of responses than anticipated. These responses, coupled with responses from ads placed in several national computer publications, placed a heavy administrative load on the PLOTALL staff. Follow-up letters were drafted and sent, together with a PLOTALL user's manual to all respondents. When time permitted, telephone contact was made with respondents showing a strong interest. This personal form of response was important in selling several packages. (One interesting point of consideration for state-supported institutions: As expected, we received several responses from "sister" state schools. It was determined that copies of the PLOTALL software package must be made available to Ohio state-supported schools upon request. No fee may be charged for software packages developed in the "Public Domain.")

DISTRIBUTION

Plans had been made to protect the integrity of the package by distributing it in compiled (object) form. Only a few small graphic interface routines were to be distributed in source form. The shortcomings of this philosophy were quickly discovered following the distribution of the first few packages. This method was very restrictive with respect to machine and compiler compatibilities. The desirability of FORTRAN source distribution became obvious.

Development of a distribution tape containing installation instructions, source code and a comprehensive set of test programs was the next step. PLOTALL Version 2.0 (which uses a FORTRAN language processor) has been successfully distributed in this form with no adverse comments. Updates and/or patches to this package are readily effected.

Consideration was given to the machine/device adaptability early in the PLOTALL development. This is not an easy goal to achieve, however, as many software vendors have found.

PLOTALL is presently being run on several models of IBM machines. Additionally, testing is being conducted on Honeywell and Burroughs hardware. Successful interface to a number of graphics devices has taken place. The difficulty with these "foreign" installations is, obviously, in the inability of the supplier to do routine, on-site testing.

The use of designated test sites can partially solve the "foreign" hardware problem. The tie to these sites must be formal, however, since timely testing requires a commitment on the part of the test site. No formal test sites have yet been established for the PLOTALL package.

SUPPORT/DEVELOPMENT

One of the most important lessons learned during the PLOTALL development, marketing and distribution effort was the critical need for in-depth planning and organization. These are the ingredients that make for not only a successful but also a less painful marketing experience.

Early in the marketing phase, the original PLOTALL author and the section manager left the university. This is not uncommon in today's software personnel market. The transition of responsibilities was smooth, but with new people came new philosophies, ideas and methods. In retrospect these new faces added an objective view to the effort. A period of critiquing and reorientation followed which had an overall positive effect on the product and the marketing effort. Without a strong organizational foundation, the entire effort might have ended at that point. Support of a package such as PLOTALL requires a management commitment to the package. This means that as long as a contract is in effect, matters concerning the package must have extremely high priority. Since good will and public relations are among the most important aspects of the marketing effort, the quality of package support must be excellent.

In the case of PLOTALL, further development was contingent on success in the marketing venture. As stated earlier, numerous requests for upgrades of the package have led to Version 2.1. This again is a management commitment with respect to priorities and workload. A large package such as PLOTALL nonetheless requires a great deal of time to develop and maintain.

Further development of the PLOTALL package is planned. As the diversity and the number of installations grow, so does the support function.

DEVELOPMENT AND MARKETING CONSIDERATION

The following outline was developed in an attempt to locate potential problem areas of software development and marketing from a university environment. The list is not intended to be a complete guide to successful marketing. Rather, it should be viewed as an aid in prompting questions and answers.

I. *Review development history*
   A. Initial goals
   B. Present status

II. *Define future developmental goals*
   A. Life of the package
   B. Development time table

III. *Review marketing options*
   A. In-house advantages
   B. In-house disadvantages
   C. Contract part of the effort
      1. Use of a publisher
      2. Use of a consultant marketing group
   D. Contract entire effort
      1. Commercial software vendor
      2. Commercial hardware vendor

IV. *The in-house marketing effort*
   A. Define marketing goals
   B. Review product protection (copyright, etc.)
   C. Contract considerations
      1. Purpose of contract
      2. Lease vs Sale
   D. Marketing Effort
      1. In-depth planning
      2. Administrative preparation
      3. Advertising
         a. define market
         b. review methods available
         c. produce advertising materials

V. *Distribution*
   A. Contract negotiation
      1. Contract charges/addenda
      2. Legal consideration with respect to sales
         a. Nonprofit organization
         b. Foreign companies
   B. Machine/Device compatability
   C. Administrative aspects of distribution
   D. Scale of distribution
      1. Number of sites which can be supported
      2. Number of different configurations which can be supported
   E. Format for distribution
   F. Patches and updates
      1. When to distribute
      2. Form of distribution
   G. Distribution of new versions

VI. *Miscellaneous considerations*
   A. Compensation for personal effort in
      1. Development
      2. Sales
   B. Legal responsibilities for accuracy
   C. Need for large-scale commitment to entire effort

The Distribution and Licensing of the Laboratory for Computer Graphics and Spatial Analysis Software: An ODYSSEY Case Study

by William G. Nisen

During the last thirteen years, the Laboratory has accumulated a moderate amount of experience with regard to the distribution and marketing of computer programs and cartographic data bases. However, due to the necessary time constraint imposed on this presentation, only licensing and distribution factors will be addressed. Please note that the following remarks will only be relevant to situations where programs are distributed in source code.

Probably the greatest obstacle to the prevention of unauthorized distribution of software is ignorance. It has been our experience that very few individuals who know that the software in question is proprietary will actually transfer the program or programs without proper authorization.

With the exception of the occasional occurrence of premeditated, unauthorized transfer, it is seldom worth while to ferret out the culprit unless the software is in the $10,000 range. Usually, this type of illegal transfer only constitutes a small percentage of the number of unauthorized copies. However, if out-and-out thievery is a major problem, serious consideration should be given to the release of the object program only or to some form of cryptographic coding.

Another major concern with regard to proprietary software is unauthorized use or access. This area has posed the most serious problem for the Laboratory. Approximately sixty percent of all the Laboratory's software has been distributed to universities for their own internal use. We do not limit use within the university setting, since we consider the entire university's educational community as a single user. The rub comes when organizations outside the university — either commercial, governmental, or other educational institutions — are allowed access to the software resident at the university's computing center. To allow access to our programs by these organizations is contrary to the nature of our Terms of Agreement, which are a very important part of our Order Form. We have requested that all University computing centers who allow outside organizations access to our programs have the organizations contact us for authorization to use our programs. This usually requires that the organization license the programs of interest from us.

The fourth major consideration in the licensing and protection of software is in a network environment. The table below summarizes the five major types of networks that predominate in the United States.

NETWORKS	
Time Sharing	**Educational Network**
One or more CPUs usually Commercial	One main center with many satellites with no CPUs
One or more sites (but usually one site per system)	
Programs can be executed, but may or may not be	**State Networks**
	Many agencies per CPU
	Usually remote processing predominates
Federal Gov't. Networks	**Miscellaneous**
Agency Specific	EDUNET
Satellites have CPUs	ARPANET

Each type demands a separate approach which can best be decided upon after preliminary negotiations have taken place.

Perhaps the most expeditious manner by which an examination of how these considerations can be addressed in a licensing agreement is to review the following draft version of the ODYSSEY licensing agreement which begins below.

ODYSSEY LICENSE AGREEMENT

Effective_____, the President and Fellows of Harvard College (hereafter "Harvard") and (hereafter "Licensee") having one of its principal offices at_____agree as follows:

Effective_____, the President and Fellows of Harvard College (hereafter "Harvard") and_____ (hereafter "Licensee") having one of its principal offices at _____ agree as follows:

ODYSSEY LICENSE AGREEMENT

ARTICLE 1.00 BACKGROUND

1.01   Harvard has carried on research and development and now has in its possession certain computer programs for use in the analysis and display of spatial data, together with valuable information and know-how concerning the contents of such programs.

1.02   Licensee wishes to acquire a nontransferable, nonexclusive license to use selected computer programs developed at Harvard.

THEREFORE, the parties agree as follows:

ARTICLE 2.00 DEFINITIONS

2.01   The following capitalized terms shall for purposes of this Agreement have the meanings set forth in the succeeding sections of this ARTICLE 2.00: .

2.02   ORDER FORM shall refer to the Program Order Form attached to this Agreement and hereby incorporated herein.

2.03   HARVARD PROGRAMS or the PROGRAMS shall mean the program packages or portions thereof designated in the Order Form, together with any corrections, improvements, replacements or associated documentation provided by Harvard and any copies, modifications, or additions to such material made by Licensee.

2.04   USE shall mean the copying of any portion of the Harvard Programs, in any machine readable form, from storage units or media into computer equipment for processing, or the reference to any portion of the Harvard Programs in printed form in support of any processing activities.

2.05   DESIGNATED CPU shall mean the particular computer central processing unit designated by type/serial number and located at the location specified in the Order Form.

2.06   AUTHORIZED USERS of the Harvard Programs shall mean only the Licensee's employees who are Using the Harvard Programs solely in connection with the Licensee's operations.

ARTICLE 3.00 LICENSE GRANT

3.01   Subject to the terms and conditions of this Agreement, Harvard hereby grants to Licensee and Licensee hereby accepts a nontransferable, nonexclusive license to Use the Harvard Programs with the Designated CPU. A separate Agreement is required for any Use of the Harvard Programs, in any machine readable form, at any location other than the Designated CPU. No right is granted for the Use of any Harvard Programs by other than Authorized Users of the Designated CPU.

3.02   Licensee agrees to confine its Use of the Harvard Programs, in any machine readable form, to the Designated CPU. Licensee agrees to use its best efforts to limit access to the Harvard Programs to Authorized Users of the Designated CPU and to permit access to the Harvard Programs to no other persons. The Harvard Programs shall not be Used on any processing unit other than the Designated CPU nor on the Designated CPU if the location of the Designated CPU is changed, unless Harvard has given its express prior written consent to such change of location.

3.03   Harvard retains all of its rights to the Harvard Programs, and Licensee agrees that it does not have and shall not claim at any time any proprietary or other rights to any of the Harvard Programs, other than the limited right to Use such Programs granted under this Agreement. Licensee further agrees that in all presentations and papers regarding the Harvard Programs or including information derived from their Use, reference shall be made to the Harvard Laboratory for Computer Graphics and Spatial Analysis as the supplier and developer of the Programs. No other public reference to the Laboratory or Program(s) name, or use of the name "Harvard" shall be made in connection with the Programs or information derived from their Use without the express prior written consent of Harvard.

3.04   Licensee may convert or copy the Harvard Programs, in machine readable form, in whole or in part, into printed or machine readable form for the Licensee's Use with the Designated CPU, provided that no more than two (2) printed copies and machine readable copies of each Harvard Program or parts thereof will be in existence at any time without the express written consent of Harvard.

3.05   Except as provided in Section 3.04 of this Agreement, Licensee agrees not to allow any other individual or organization to copy any portion of the Harvard Programs in any form for any purpose whatsoever without the express prior written consent of Harvard. All copies of each Harvard Program, or any portion thereof, whether in printed or machine readable form, shall be deemed part of the Harvard Programs, and Licensee agrees to reproduce and include Harvard's notice of its rights therein on all copies, or partial copies, made under this Agreement.

3.06   Licensee may modify the Harvard Programs, in machine readable form, to meet the particular requirements of the Licensee's own use, and to this extent may merge the Harvard Program into other program material to form an updated work. Any modifications or additions to the Harvard Programs shall be deemed part of the Harvard Programs and shall be used only with the Designated CPU. Upon termination of the license to Use the Harvard Programs for any reason, said Programs and any associated material shall be removed from such updated work and shall either be returned to Harvard or destroyed in accordance with Section 8.02.

3.07   Licensee agrees not to market any data processing service that makes Use of the Harvard Programs unless express prior written consent has been obtained from Harvard.

3.08   Neither this Agreement nor any rights hereunder, in whole or in part, shall be assignable or otherwise transferable.

ARTICLE 4.00 LICENSE TERM AND FEES: PROGRAM DELIVERY AND CANCELLATION

4.01   Each license granted herein shall run for one (1) year from the effective data of this Agreement and is renewable for subsequent periods of one (1) year upon Harvard's acceptance of written notice from Licensee of Licensee's intention to extend such license for each additional year. Licensee shall provide Harvard with such written notice at least ninety (90) days prior to the end of the then current period of this Agreement. Harvard in its sole discretion may grant the additional term. Harvard shall notify the Licensee of its acceptance in writing at least thirty (30) days prior to the end of the then current period.

4.02   In consideration for the Use of the Harvard Programs and Harvard's other services under this Agreement, Licensee agrees to pay to Harvard_____dollars ($____) within thirty days from the effective date of this Agreement. Upon delivery of all capabilities described in the Order Form, Licensee shall pay to Harvard the remaining   dollars ($) of the initial license fee.

4.03   All license fees shall be paid to Harvard without reduction for any tariffs, duties, or taxes imposed or levied by any government or governmental agency. Licensee shall be liable for payment of all such taxes, however designated, levied or based on the Harvard Programs, their Use, or on this Agreement, including without limitation thereto, state or local sales, use, and personal property taxes.

4.04   Delivery of the Harvard Programs shall be made in accordance with the delivery schedule set forth in the Order Form. All computer programs and other items recorded on magnetic tape or punched cards will be shipped to Licensee by best method, postage and handling prepaid. Documentation materials normally provided with such Programs will be included as part of the shipment. All other materials shall be sent as third class mail, printed matter.

4.05   Licensee may cancel delivery of any portion of the Harvard Programs not yet received and will be liable for the license fee due on or before the date of cancellation.

ARTICLE 5.00 PROPRIETARY AND CONFIDENTIAL NATURE OF PROGRAMS

5.01   Licensee acknowledges that the ideas expressed in the Harvard Programs and the Harvard Programs per se are confidential proprietary information belonging to Harvard and agrees that it will take all reasonable steps to maintain their confidential proprietary nature. Licensee further agrees to use the same degree of care to avoid dissemination or discharge of the Harvard Programs or copies thereof to persons, other than Authorized Users, as Licensee uses with respect to its own proprietary computer programs of like importance, provided that such dissemination or disclosure may be made upon Licensee's receipt of Harvard's prior written consent. Licensee shall also notify each Authorized User to whom any disclosure is made, that the Harvard Programs are confidential and proprietary and are subject to this nondisclosure provision.

5.02   The obligation of Licensee under 5.01 shall survive and continue ten (10) years after any termination of rights under this Agreement; however, such obligation shall not extend to the Harvard Programs or related information which: (1) are in the public domain at the time of disclosure; (2) have been independently discovered by Licensee prior to disclosure; (3) later becomes available to the general public by reason of acts not attributable to the breach by Licensee of its obligations under this Agreement; or (4) are the same or similar in substance to information furnished by Harvard to third parties without proprietary restrictions.

5.03   In the event of a willful violation by Licensee of any of the terms and conditions of this ARTICLE 5.00, Harvard shall have the right, in addition to other rights and remedies available to it, to injunctive relief without posting a bond, it being specifically acknowledged that other remedies are inadequate.

ARTICLE 6.00 PROGRAM SUPPORT AND MAINTENANCE

6.01   Harvard shall provide corrections and updated documentation of the Harvard Programs if and when developed during the license term at no additional cost to Licensee. Har-

vard will be available for written or telecommunication consultation to remedy errors which the Licensee can clearly document as existing in the original source code of the Harvard Programs. Under no circumstances will any individual associated with Harvard be required to perform on-site services, unless a fee for service arrangement has been negotiated in advance. Neither the execution of this Agreement nor any action taken by Harvard in connection with the Harvard Programs shall be construed as an undertaking by Harvard to furnish Licensee with any assistance except as set forth in this ARTICLE 6.00, nor shall any action be taken by Harvard impose upon Harvard any obligations to furnish such assistance.

6.02   Harvard agrees to provide any updated versions of the Harvard Programs, in machine readable form, developed during the term of this Agreement to the Licensee for the then current price of the Program replacement tapes. Replacement tapes may be obtained at any time during the term of this Agreement by giving written assurances that the current source tape is no longer in satisfactory condition and will be destroyed upon receipt of the replacement tape. All such updated versions or replacement tapes shall be considered a part of the Harvard Programs and subject to the terms and conditions of the Agreement.

ARTICLE 7.00 REPRESENTATIONS AND WARRANTIES

7.01   Harvard and Licensee agree that the contents of the Harvard Programs are fully defined by the distribution tapes and associated documentation designated in the Order Form. Harvard and Licensee further agree that there are no other understandings, agreements, warranties, or representations, express or implied, between the parties with respect to, or relating to, the contents of the Harvard Programs.

7.02   In particular, but without limitation, HARVARD MAKES NO REPRESENTATION OR WARRANTY OF MERCHANTABILITY, OR OF FITNESS FOR ANY SPECIFIC PURPOSE OR USE.

7.03   To the best of Harvard's knowledge and belief, none of the Harvard programs delivered hereunder has been copied from another source.

ARTICLE 8.00 TERMINATION AND BREACH

8.01   Upon the happening of any of the following events there shall be deemed to be a breach of the terms of this Agreement, and Harvard and Licensee shall have the right, in addition to any other remedies that they may have to terminate all rights granted hereunder by written notice to the other party:

   (1) failure of the Licensee to pay in full, in accordance with the terms of this Agreement, any license fees or other charges included herein;

   (2) violation of any of the terms and conditions of this Agreement by either party, in continuation of such violation for thirty (30) days after receipt of written notice thereof;

   (3) either party becoming insolvent, making an assignment for the benefit of creditors, filing a petition for bankruptcy or other action by either party by their respective creditors, provided that in the event of such acts on behalf of Harvard, the Har-

vard Programs in the possession of the Licensee pursuant to the Agreement shall become the property of the Licensee.

8.02   Except as provided in section 8.01(3), upon the termination of this Agreement, whether pursuant to the terms hereof, or by allowing it to lapse, or otherwise, the Licensee shall remove the Harvard Programs and any material associated therewith from any location in which the Programs are being Used and within ten (10) days either return such material to Harvard or destroy it and deliver a written certification to that effect.

8.03   Any questions concerning the interpretation of this Agreement or dispute arising out of or connected with this Agreement shall be discussed by Licensee and Harvard, and if the parties are unable to resolve any such questions or disputes by discussion, such questions or disputes shall be submitted to arbitration in accordance with the then pertaining rules for commercial arbitration of the American Arbitration Association. Judgment on the award in any such arbitration may be entered in any court of competent jurisdiction. Any such arbitration shall be held in Boston, Massachusetts.

ARTICLE 9.00 LIMITATION OF LIABILITY

9.01   Harvard shall not, by reason of termination or non-renewal of this Agreement, be liable to the Licensee for compensation, reimbursement, or damages due to expenditures, investments, leases, or commitments in connection with the business or goodwill of the Licensee or otherwise.

ARTICLE 10.00 MISCELLANEOUS PROVISIONS

10.01   If any provision or provisions of this Agreement shall be held invalid, illegal, or unenforceable, and such provision(s) does not constitute a substantial part of this Agreement such that its deletion from this Agreement would defeat the purpose of this Agreement, then the validity, legality, or enforceability of the remaining provision shall not be affected.

10.02   This Agreement shall be governed by and construed in accordance with the laws of the Commonwealth of Massachusetts.

10.03   No delay or failure of Harvard or Licensee in exercising any rights hereunder shall be deemed to constitute a waiver of such rights or any other rights hereunder.

10.04   All notices or communications called for under this Agreement shall be made to the individuals designated in the Order Form. Terms and provisions of this Agreement shall not be modified, amended, altered or changed except by written agreement signed by the parties hereto.

10.05   Terms of this Program License Agreement shall be kept in confidence by Licensee.

10.06   This Agreement constitutes the entire Agreement between Licensee and Harvard, and no representation, agreement, or promise of any officer, employee, or agent of Licensee or Harvard not set forth herein shall in any way affect the obligations of the parties hereto as herein set forth.

A Case Study of the Use of Computer Mapping Evaluating Election Returns and Relevant Socio-Economic Indicators

by John K. Wildgen

THE SEAMY SIDE OF DEMOCRACY

"Few would contest the proposition that among its professional politicians of the past two decades Louisiana has had more men who have been in jail, or who should have been, than any other American state," was the way V.O. Key, Jr. opened his chapter on Louisiana in the 1949 classic, **Southern Politics.**[1] In the more than thirty years since Key wrote those lines, nothing in Louisiana's conduct of the public's business could serve to justify the slightest modification of his assessment.

It may be that the citizenry of other states (Illinois, New Jersey, and Massachusetts) would contest Louisiana's primacy in corruption and try to match anecdote for anecdote, scandal for scandal. Yet there is an undeniable flavor to political corruption in Louisiana that makes it *sui generis.* For example, it may be the case that politicians outside of Louisiana have taken tens of thousands of dollars in illegal contributions. But how many of them have, when caught at it, claimed that the illegality consisted in offering the money, not in taking it? Furthermore, how many of them have subsequently convinced both law-enforcement authorities and the electorate of the rectitude of that position?

It is clearly impossible to settle arguments about how corrupt a given system's politics are. But in Louisiana, things are pretty bad — so bad that the population thinks the state is the worst, and curiously, takes a perverse pride in that perception.

This pride in the state's low level of official morality has had unusual and difficult consequences for instruction in political science. The atmosphere of corruption creates a kind of smokescreen in front of other aspects of state politics that are of central theoretical interest to political scientists specializing in either American or comparative politics. It is, to put it bluntly, hard for academics to distract students of politics from the stranger-than-fiction financial and sexual exploits of the state's leadership.

The mixture of political phenomena that lurks behind the smokescreen of corruption is full of paradox. At the most obvious level to observers lie the "third-world" aspects of life in the state, such as low industrialization, specialized agriculture, illiteracy, class conflict, religious bigotry, and racism. But alongside these barriers to modernization there is a largely hidden political infrastructure of long duration that for nearly a century has been state-of-the-art. For example, Louisiana has a

tradition of high political participation or electoral turnout. Indeed, in the 1890's Louisiana was attempting to reverse adult male suffrage nearly a quarter of a century before nations like Britain and Italy achieved it in the first place. Under the guise of one-party rule, Louisiana developed in the 1930's a bifactional system that amounted to model two-party competition along British lines of the 50's and 60's. In 1959-1960 Louisiana experienced a major shift in the geographical distribution of its votes — a critical election. In 1971 the state underwent massive dealignment as all signs of bifactionalism disappeared in a welter of multifactional voting.

These aspects of Louisiana politics: political mobilization, bifactionalism, critical elections, and dealignment — all have parallels in American national politics and in the politics of other advanced post-industrial nations. But, given the atmosphere in the state, it is not hard to visualize three difficulties in offering instruction on Louisiana politics: (1) avoiding a fixation on corruption; (2) explaining the meaning of concepts like mobilization, bifactionalism, critical elections, and dealignments in the context of Louisiana; and (3) explaining why such completely modern political phenomena occur, often unexpectedly early, in a state that has little else modern except politics.

Each of these difficulties requires a different resource or capability. Avoiding discussions of corruption in Louisiana political science classes requires disciplined students and a very disciplined instructor. Explaining the paradox of modern politics in a pre-modern society and economy requires a talented political scientist. Explaining mobilization, bifactionalism, critical elections, and dealignment requires good computer graphics. This report concentrates on this last point.

MAPPING MODERN POLITICS: METHODS AND RESOURCES

Descriptive or Analytical Objectives

The kinds of modern political phenomena discussed above are rather like some fundamental concepts in physics in that electorates are like mass, bifactionalism (partisanship) is like energy, critical elections are like motion, and dealignment is like entropy. Too much should not be made of these analogies, of course, but the point is that political scientists and physicists are both quite inured to dealing with concepts that are sometimes easier to illustrate than define.

Yet the problem of illustration of concepts and phenomena is hardly trivial for either a physical or political endeavor. For political scientists, though, there is a special problem in the fact

---

[1]There is an extensive literature on Louisiana politics. Aside from Key (1949), some of the most notable monographic works include Williams (1969), Sindler (1956), and Howard (1957, 1971). For a less formal introduction, Liebling (1970) is universally recommended.

that members of that discipline are uncomfortable with the science in political science. To a political scientist, it is important to be analytical rather than descriptive. This puts computer graphics, especially cartography, at a great psychological disadvantage in political science just because maps are such excellent descriptive devices. It is supposed, often unconsciously, that anything that is good for description cannot be good for analysis. Things that are analytical should have "analysis" in their names — things like regression analysis, factor analysis, and discriminant analysis. But it is often forgotten that these useful statistical procedures are largely descriptive rather than inferential. Moreover, they are all spatially oriented. Despite that, political science has generally ignored the analytical utility of automated cartography.[2] This is particularly unfortunate in the field of undergraduate instruction since important concepts are simply harder to convey to undergraduates by means of equations than by graphic illustration. Somehow, though, one gets the impression in dealing with political scientists that Pearson's *r* is more "scientific" than an associated scattergram. Further, if the data are concerned with a geographical unit, the scattergram is often perceived as being more "scientific" than a map.

Graphics Resources At a Medium-Size University

Given a commitment to employ computer graphics and cartography to problems in political science, how does one go about doing it? At the University of New Orleans, the gathering of resources was an outcome of a collection of fortunate coincidences unlikely to occur elsewhere. But the necessary equipment and software do have a total cost that is not cheap: the basic hardware used in this report was a Tektronix 4014-1 display terminal (about $11,650) and an accompanying Tektronix 4631 Hard Copy Unit (about $4,295). The terminal was connected by a 2400 baud line to a Decsystem-10 with 256K words of (36 bit) main storage and assorted disk and tape drives.

Software and data bases were of two types. The statistical data bases on Louisiana's electoral history were gathered and encoded under the supervision of the author from state records. Socio-economic data were gathered largely, but not exclusively, from files maintained at the Inter-University Consortium for Political and Social Research (ICPSR). The geographic data base — outlines of Louisiana's sixty-four parishes — came in DIME format from Harvard's Laboratory for Computer Graphics and Spatial Analysis. In terms of software, statistical processing of the data was done by SPSS, Western Michigan's STP, and Pennsylvania State's MINITAB. Cartographic software employed included Harvard's POLYVRT to process the GBF for Louisiana and INPOM to generate conformant maps.

With the exception of INPOM, all software and data were acquired through routine public or proprietary channels. In the next subsection some comments about INPOM are offered, along with some observations about difficulties with the array of graphic tools assembled at U.N.O.

---

[2]This is not intended as a blanket indictment but as a description of what happens most of the time. For example, the Inter-University Consortium for Political and Social Research (1977) issues a catalogue of data and services that is roughly 450 pages in extent. The organization is probably the most widely representative body in existence of the interests of contemporary political scientists. So far the ICPSR has simply not had appreciable demand from its constituency to get involved in graphics software or data bases.

Problems Encountered

It is our current understanding that, outside of Harvard, INPOM is operating only at U.N.O. We are quite happy with it — we suspect more happy than Harvard, which promises a better replacement soon. Be that as it may, we can certainly vouch for convenience and speed in terminal cartography. But there are two potential problems to beware of: one hardware, the other software. In terms of hardware problems, it is essential to locate terminals where they have access to high speed lines. Mapping programs transmit a lot of signal and 2400 baud is a frugal rate for images of moderate complexity. In terms of software, we were quickly disabused of the notion that transporting a program from one DEC-10 to another would be routine. Our processor, operating system, and FORTRAN compiler were not identical to Harvard's system. So, some systems work was necessary to bring INPOM up. In short, the universal difficulty in computer science of software transportability shows up in graphics as well, in spades.

The terminal employed created no great problems aside from critical focusing of the Z-axis. On the other hand, the 4631 Hard Copy Unit does present a difficulty because of the thermal paper it uses. In New Orleans' fetid climate, the paper degrades quickly when environmentally and cost-conscious deans cut air-conditioning off in mid-afternoon. Try to get your own window unit or a sympathetic dean.

GRAPHICS EXAMPLES

How Maps were generated

For those involved in computer mapping, one of the most frustrating problems is the wide variety of base files used to describe boundaries and spaces. Leaving aside the relative merits of grid versus polygon approaches in general, I think it is reasonable to argue that, for the most part, mapping aggregate political phenomena (county voting percentages, for example) is a polygon rather than a grid problem. Now, since the geographical units dealt with were Louisiana's civil parishes, the most obvious existing digitization of parish boundaries was found in county-level DIME files. On the other hand, INPOM assumes data in its own format which is somewhat different in philosophy from DIME files.

Given this difficulty, the first step was the creation of suitable INPOM input from the DIME records. This was done via POLYVRT (Laboratory for Computer Graphics, 1971). Since the use of POLYVRT was somewhat complicated and since it is likely to be around for some time, attention will be given to describing what was involved. POLYVRT is designed to translate existing base files into other base file formats. The tape comes with a DIME file of U.S.A. states and counties and a World Data Bank in chain format with boundaries of nations encoded. The tape comes in EBCDIC. Thus, one of our first tasks was one of translating the DIME file from EBCDIC to ASCII for use on a DEC-10. The utility employed for that purpose was CHANGE, well known at DEC installations.

The preparation of the POLYVRT program itself required some cheerfully given advice from Nicholas Chrisman at the Harvard Laboratory since the POLYVRT manual makes no mention of the program's ability to write INPOM files. There are two of them, one a chain file, the other a polygon file. In the coding for POLYVRT, the two files are written on the same channel, one after the other. On a DEC-10 this creates a difficulty unless a sequential output device such as magnetic tape is used.

POLYVRT itself is a reasonably simple program to run due to the clarity of manual and the built-in defaults. Our own particular application was quite straightforward and did not require much more than simple-minded use of the program. The only annoying difficulty about the program from the point of view of a terminal-oriented system was the need to adhere to a column-specific format. Figure 1 is a copy of the program used to generate the data input to INPOM. Execution time was about eight minutes.

The next step involved using the files output from POLYVRT as input to INPOM to create an INPOM GBF. Since INPOM will be replaced in the near future, there is no reason to go into any detail regarding its own idiosyncracies. But there are some features of it that are likely to show up in one form or another in substitutes for it. One well-known problem is that of class intervals (Monmonier, 1972). Another is the question of shading. In this particular case, neither problem was solved in any final sense. Class intervals were set to quartiles simply because previous political science literature on Louisiana has resorted to them. It is a fortunate coincidence that the state has sixty-four parishes. The selection of shading is a question of both aesthetics and speed. Vectors are usually drawn much more rapidly than are characters such as "x" or any other suitable literal. In general, students seem to be more comfortable with shadings that are based on both characters and hatchings.

What Emerged

In this section, a couple of examples will be presented that are suggestive of both the descriptive and analytical utility of statistical mapping of theoretically interesting political phenomena. Each example will contrast mapped results with correlation coefficients and scattergrams.

The first example takes up the notion of political mobilization or how people are induced to register to vote. Louisiana's minority problem is well known but there is a general notion in social science that minorities anywhere are not *uniformly* oppressed. It is suspected that smaller minorities fare better because they are neither so obvious nor so threatening. In terms of voter registration, this implies that, everything else being equal, Blacks should have found it easier to register in

Louisiana parishes where there were relatively fewer blacks.

In Example 1, we begin with a standard statistical approach to the problem in which an index of black toleration (in terms of voter registration) is plotted and correlated with per cent of black population. The index is computed for 1956, at the beginning of Earl Long's second term as governor. The population data came from the 1960 census. The scattergram is not an overwhelmingly suggestive device here, though an experienced statistician does not expect a product-moment correlation of -0.58 to fit a line closely. On the other hand, the two maps drawn from the same data make it very clear that blacks are rare in western Louisiana and, as the hypothesis predicts, tolerated at registration time. Of course, there are several areas that do not conform to the implied model of toleration. But when the outliers are located on a map rather than on a scattergram, even statistically naive students perceive both the rough fit of the model and the need to find additional explanations.

Example 1's analytical utility lies largely in its ability to raise more questions than it answered. But in Example 2 we saw an instance where resort to mapping was not just a marginally better kind of descriptive statistic, but led to inferences largely beyond the scope of routine statistics.

The substantive example is a case of critical election, a major realignment of state voting patterns (Key, 1955). In the previous example we were expecting a non-zero correlation. In the case of realignment, in contrast, the expectation is an *r* of zero and a random-patterned scattergram. The first leaf of the example confirms the expectation. Of course, when students see this sort of thing they note it and understandably fail to comment. It is hard to discuss zero correlations at length with any enthusiasm.

The maps of the two elections, Earl Long's 1956 vote and Jimmie Davis' 1960 vote, are markedly different in appearance — only what the correlation and scattergram would lead us to expect. But Davis' voting pattern shows the emergence of a distinct regional, north-south, pattern. This is something not even hinted at in the statistical output.

DISCUSSION OF RESULTS

Reaction of Students and Colleagues

It was in the early years of the 1970's that undergraduate education and computers met at U.N.O. The result was the same as elsewhere for the most part: Students, by and large,

```
A-INPUT    DIMECO
B-SELECT
IF         STATE      EQ         22
END
C-READ
UNTIL      RECORD     EQ         35110
END
E-MANIPULATE
PROJECT    ALBERS
           MERIDIAN   -90.00
           PARALLELS  28.5       33.2
END
F-GENERAL
END
G-DRAW     ANNOTATED                         20.0     20.0
G-OUTPUT   INPOM                                  25.
END
Z-FINISH
```

Figure 1.

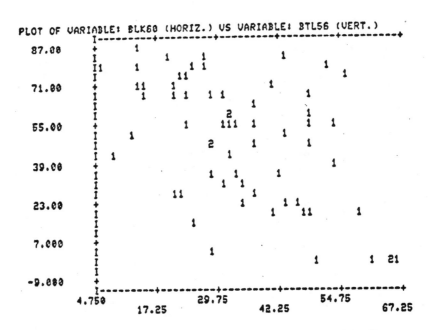

WHICH COMMAND? CORR

***** CORRELATION MATRIX *****

```
VAR.
BTLSS    1.0000
BLKSO   -0.5311   1.0000

         BTLSS    BLKSO
```

WHICH COMMAND?

Exhibit 1. A.

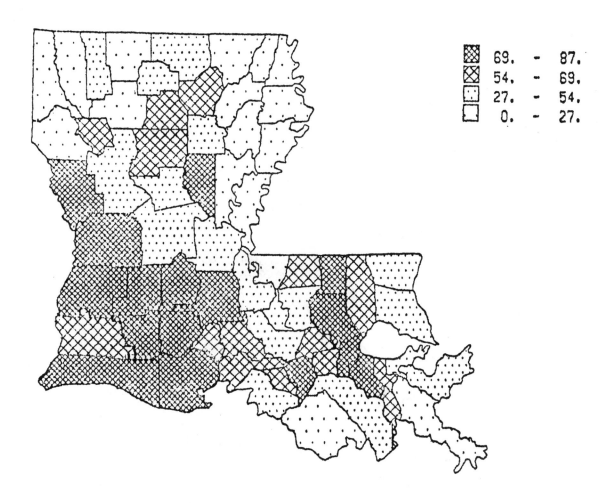

Exhibit 1. B. Toleration of Black Registration — 1956

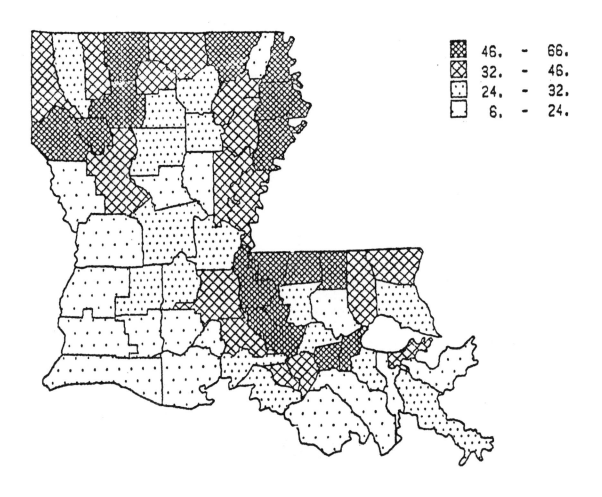

Exhibit 1. C. % of Population Black — 1960

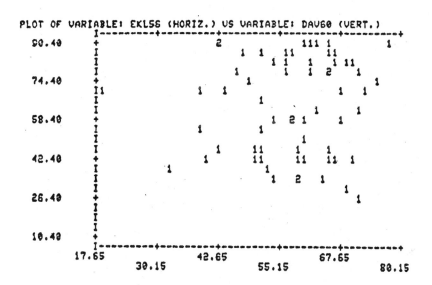

PLOT OF VARIABLE: EKL5S (HORIZ.) VS VARIABLE: DAV6O (VERT.)

WHICH COMMAND? CORR

***** CORRELATION MATRIX *****

VAR.		
EKL5S	1.0000	
DAV6O	0.0657	1.0000
	EKL5S	DAV6O

WHICH COMMAND?

Figure 2.

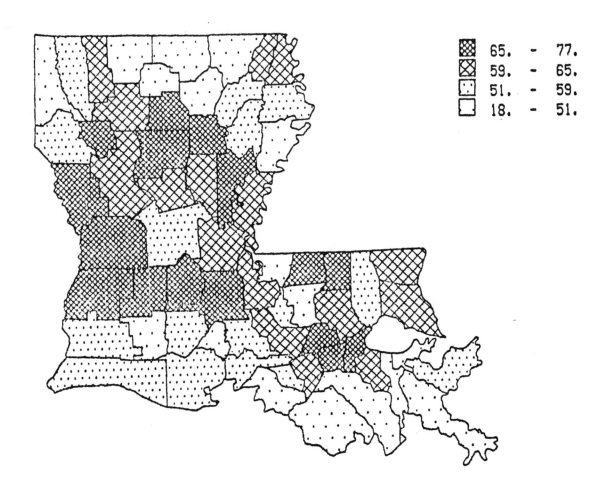

Exhibit 2. A. Vote for Earl Long — 1956

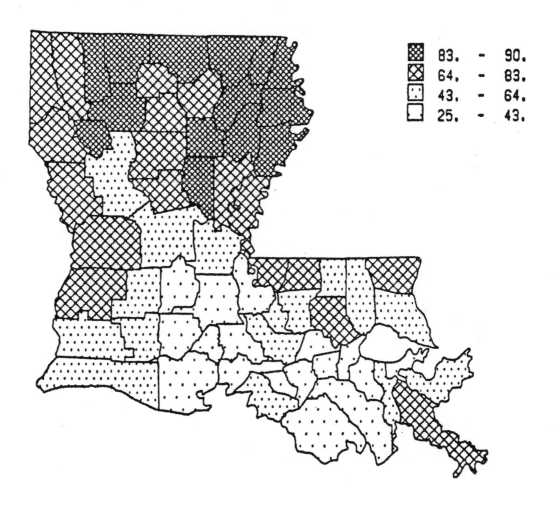

Exhibit 2. C. Vote for Jimmie Davis — 1960

were petrified. Some were angry and saw computers as another example of everything they saw wrong about America. A few, on the other hand, became addicted. Computer mapping, in contrast, has had a more sanguine reception because it does make statistical arguments clearer. After all, maps are pictures, not numbers. Furthermore, computer mapping is still rarely taught. Most social science departments have in-house statisticians and computer types who make students calculate and write programs in packages like SPSS, but very few involve students in mapping. The software is much more complex and the equipment too rare and expensive to turn students loose on it. So, mapping is an aid, not a threat, to the bulk of them.

But what kind of aid is it? In teaching a course in politics, regardless of the geographical area concerned, maps speed understanding of concepts that are geographically dispersed. In certain instances, they do not merely sugar-coat numerical examples but are superior to them. This, however, can be the case only when the instructor avoids an atlas mentality. After all, the batch mapping of data is nothing more than the generation of peculiar histograms. Routine statistical procedures instruct very well, and while the examples chosen here make a case for the use of maps, it is not being suggested that they replace other aids.

The reaction of other faculty has also been mostly sanguine. The only difficulty is that demand can quickly exceed the ability to render service. Geographers seem to understand this. Besides, many of them have no interest in cartography anyway. In general, the major disappointment encountered so far in dealing with faculty has been the fact that it is much easier to display a GBF than create one. We are currently in the embarrassing situation of being able to map all American states and counties, and any country in the world, while we have not succeeded in getting a good rendering of local police districts.

Future Uses

The case study reported here had promising results. Recall that the basic instructional problem was to distract students from an anecdotal view of state politics and get them to dwell for awhile on some aspects of the state's political structure that exist rather independently of day-to-day boodling by public officials. The results were successful enough to make possible funding for more extensive work in this field and a subsequent publication of results. The results will be less an essay in political geography than a more modest geography of politics in Louisiana. Themes to be discussed, in addition to the ones mentioned, will include analyses of turnout, the insulation of state politics from national patterns of partisanship, and policy outputs.

In a sense, the kind of technology we will be in a position to transfer will not be graphic hardware or software; indeed, we are mere consumers of those products. Instead, we are concentrating on developing suitable and critical applications to be used as analytical tools of political science.

BIBLIOGRAPHY

HOWARD, PERRY H. (1957) *Political Tendencies in Louisiana 1812-1952.* Baton Rouge: LSU Press.

HOWARD, PERRY H. (1971) *Political Tendencies in Louisiana,* Revised and Expanded Edition. Baton Rouge: LSU Press.

INTER-UNIVERSITY CONSORTIUM FOR POLITICAL AND SOCIAL RESEARCH (1977) *Guide of Resources and Services 1977-1978.* Ann Arbor: ICPSR.

KEY. V.O., JR. (1949) *Southern Politics.* New York: Knopf.

KEY, V.O., JR. (1955) "A Theory of Critical Elections." *Journal of Politics* 17 (February): 3-18.

LABORATORY FOR COMPUTER GRAPHICS AND SPATIAL ANALYSIS (1971). *POLYVRT Users Manual.* Cambridge: Laboratory for Computer Graphics and Spatial Analysis.

LIEBLING, A.J. (1970) *The Earl of Louisiana,* Baton Rouge: LSU Press.

MONMONIER, MARK S. (1972) "Contiguity-Biased Class-Interval Selection: A Method for Simplifying Patterns on Statistical Maps." *Geographical Review* 62 (April): 203-228.

SINDLER, ALLAN P. (1956) *Huey Long's Louisiana.* Baltimore: Johns Hopkins Press.

WILLIAMS, T. HARRY (1969) *Huey Long.* New York: Knopf.